EMERGENCY ALERT AND WARNING SYSTEMS

CURRENT KNOWLEDGE AND FUTURE RESEARCH DIRECTIONS

Committee on the Future of
Emergency Alert and Warning Systems:
Research Directions

Computer Science and Telecommunications Board

Division on Engineering and Physical Sciences

A Consensus Study Report of

The National Academies of
SCIENCES • ENGINEERING • MEDICINE

THE NATIONAL ACADEMIES PRESS
Washington, DC
www.nap.edu

THE NATIONAL ACADEMIES PRESS   500 Fifth Street, NW   Washington, DC 20001

Support for this project was provided by the Department of Homeland Security Science and Technology Directorate, with assistance from the Department of Health and Human Services under award number HHSP233201400020B. Any opinions, findings, conclusions, or recommendations expressed in this publication do not necessarily reflect the views of any organization or agency that provided support for the project.

International Standard Book Number-13:   978-0-309-46737-7
International Standard Book Number-10:   0-309-46737-3
Digital Object Identifier: https://doi.org/10.17226/24935

This report is available from:

Computer Science and Telecommunications Board
National Research Council
500 Fifth Street, NW
Washington, DC 20001

Additional copies of this publication are available for sale from the National Academies Press, 500 Fifth Street, NW, Keck 360, Washington, DC 20001; (800) 624-6242 or (202) 334-3313; http://www.nap.edu.

Copyright 2018 by the National Academy of Sciences. All rights reserved.

Printed in the United States of America

Suggested citation: National Academies of Sciences, Engineering, and Medicine. 2018. *Emergency Alert and Warning Systems: Current Knowledge and Future Research Directions*. The National Academies Press, Washington, DC. doi: https://doi.org/10.17226/24935.

The National Academies of
SCIENCES · ENGINEERING · MEDICINE

The **National Academy of Sciences** was established in 1863 by an Act of Congress, signed by President Lincoln, as a private, nongovernmental institution to advise the nation on issues related to science and technology. Members are elected by their peers for outstanding contributions to research. Dr. Marcia McNutt is president.

The **National Academy of Engineering** was established in 1964 under the charter of the National Academy of Sciences to bring the practices of engineering to advising the nation. Members are elected by their peers for extraordinary contributions to engineering. Dr. C. D. Mote, Jr., is president.

The **National Academy of Medicine** (formerly the Institute of Medicine) was established in 1970 under the charter of the National Academy of Sciences to advise the nation on medical and health issues. Members are elected by their peers for distinguished contributions to medicine and health. Dr. Victor J. Dzau is president.

The three Academies work together as the **National Academies of Sciences, Engineering, and Medicine** to provide independent, objective analysis and advice to the nation and conduct other activities to solve complex problems and inform public policy decisions. The National Academies also encourage education and research, recognize outstanding contributions to knowledge, and increase public understanding in matters of science, engineering, and medicine.

Learn more about the National Academies of Sciences, Engineering, and Medicine at **www.nationalacademies.org**.

The National Academies of
SCIENCES · ENGINEERING · MEDICINE

Consensus Study Reports published by the National Academies of Sciences, Engineering, and Medicine document the evidence-based consensus on the study's statement of task by an authoring committee of experts. Reports typically include findings, conclusions, and recommendations based on information gathered by the committee and the committee's deliberations. Each report has been subjected to a rigorous and independent peer-review process and it represents the position of the National Academies on the statement of task.

Proceedings published by the National Academies of Sciences, Engineering, and Medicine chronicle the presentations and discussions at a workshop, symposium, or other event convened by the National Academies. The statements and opinions contained in proceedings are those of the participants and are not endorsed by other participants, the planning committee, or the National Academies.

For information about other products and activities of the National Academies, please visit www.nationalacademies.org/about/whatwedo.

COMMITTEE ON THE FUTURE OF EMERGENCY ALERT AND WARNING SYSTEMS: RESEARCH DIRECTIONS

RAMESH RAO, University of California San Diego, *Chair*
JAMES CAVERLEE, Texas A&M University
ROOP DAVE, Information Technology Research Academy, New Delhi
EVE GRUNTFEST, California Polytechnic State University
BROOKE LIU, University of Maryland
LESLIE LUKE, Los Angeles County Office of Emergency Management
DENNIS MILETI, University of Colorado, Boulder
NAMBIRAJAN SESHADRI, Broadcom Corporation (retired)
DOUGLAS SICKER, Carnegie Mellon University
KATE STARBIRD, University of Washington
CHARLES L. WERNER, ParadeRest and Commonwealth of Virginia

Staff

JON EISENBERG, Director, Computer Science and Telecommunications Board
VIRGINIA BACON TALATI, Program Officer
KATIRIA ORTIZ, Research Associate
JANEL DEAR, Senior Program Assistant

COMPUTER SCIENCE AND TELECOMMUNICATIONS BOARD

FARNAM JAHANIAN, Carnegie Mellon University, *Chair*
LUIZ BARROSO, Google, Inc.
STEVEN M. BELLOVIN, NAE,[1] Columbia University
ROBERT F. BRAMMER, Brammer Technology, LLC
DAVID CULLER, NAE, University of California, Berkeley
EDWARD FRANK, Cloud Parity, Inc.
LAURA HAAS, NAE, University of Massachusetts, Amherst
MARK HOROWITZ, NAE, Stanford University
ERIC HORVITZ, NAE, Microsoft
VIJAY KUMAR, NAE, University of Pennsylvania
BETH MYNATT, Georgia Institute of Technology
CRAIG PARTRIDGE, Raytheon BBN Technologies
DANIELA RUS, NAE, Massachusetts Institute of Technology
FRED B. SCHNEIDER, NAE, Cornell University
MARGO SELTZER, Harvard University
MOSHE VARDI, NAS[2]/NAE, Rice University
KATHERINE YELICK, NAE, University of California, Berkeley

Staff

JON EISENBERG, Senior Director
LYNETTE I. MILLETT, Associate Director, CSTB, and Director, Cyber
    Resilience Forum

SHENAE BRADLEY, Administrative Assistant
EMILY GRUMBLING, Program Officer
RENEE HAWKINS, Financial and Administrative Manager
KATIRIA ORTIZ, Associate Program Officer
JANKI PATEL, Senior Program Assistant

For more information on CSTB, see its website at
http://www.cstb.org, write to CSTB at
National Academies of Sciences, Engineering and Medicine,
500 Fifth Street, NW, Washington, DC 20001, call (202) 334-2605, or
email the CSTB at cstb@nas.edu.

---

[1] Member, National Academy of Engineering.
[2] Member, National Academy of Sciences.

Preface

More than 60 years of research on disaster response has yielded many insights about how people respond to information indicating that they are at risk and under what circumstances they are most likely to take appropriate protective action. This work was largely done in the context of traditional media. The landscape for public alerts and warnings changed with the introduction of the Internet, mobile phones, and their applications, such as social media. Following a series of natural disasters, including Hurricane Katrina, that revealed shortcomings in the nation's ability to effectively alert populations at risk, Congress passed the Warning, Alert, and Response Network (WARN) Act in 2006. This legislation encouraged the adoption of much newer technologies, including the dissemination of alerts and warning messages via mobile devices, which previous alerting technologies did not reach.

Less is known about how the use of new technologies for message dissemination and receipt changes the public response or alters how public safety officials can best employ the alerting capabilities. For example, fairly little is known about how to maximize the effectiveness of messages whose content is limited by technology constraints or policy decisions, or how best to make use of alerts and warnings in today's information-rich environments. Additionally, formal study of the use of social media in disasters has been limited, and there are many outstanding questions, including how they can be used by government officials to both alert the public and gain situational awareness, the challenges and opportunities additional input from citizens provides, the associated

> **BOX P.1**
> **Statement of Task**
>
> An ad hoc committee will review current knowledge about how to effectively deploy and use emergency alert and warning systems and explore related future computing, engineering, and social science research needs. A workshop will be convened to capture results from recent research to include work sponsored by the Department of Homeland Security's Science and Technology Directorate and to foster dialogue among technologists, social science researchers, and emergency managers. The study committee's report will summarize results from the DHS research, provide an overview of current knowledge about emergency alerts and warnings and their relationship to citizen interactions and information needs, and set forth an interdisciplinary agenda for research that will highlight gaps and future needs.

safety and privacy risks, and strategies for coping with rumors and also false information.

Research, including recent work sponsored by the Department of Homeland Security (DHS), has provided some insight into these issues. Additionally, the National Academies had previously convened three workshops under DHS sponsorship, one focusing on alerting via cell phones, one considering the use of social media, and one examining how to geographically target alerts and warnings. As part of this study, workshops were convened on August 9-10, 2016, and September 1, 2016. Workshop participants included DHS-supported researchers and other experts in disaster sociology, emergency response, and technologies. Additional briefings were held on November 1-2, 2016, January 26-27 2017, and March 23, 2017 (Appendix C provides a list of briefings received).

This report reviews results from DHS-sponsored research (Appendix B includes summaries of this work), the Academies workshops, and other sociotechnical research on the public response to alerts and warnings. Building on that review, the committee sets forth a research agenda that highlights areas where future research should be focused. (Box P.1 contains the full statement of task.)

As the committee was wrapping up its work, the nation experienced a series of major natural disasters, with devastation from hurricanes Harvey, Irma, and Maria, and the October 1, 2017, shootings on the Las Vegas Strip. Each of these events was a sober reminder of the impacts of disasters on our communities and the important role that timely and effective communication with the public plays in responding to such events. Early reports on the October 2017 California wildfires further underscore the importance of public alerting and potential benefits of enhancing the reach and effectiveness of the Wireless Emergency Alerts system, which allows public officials to deliver alerts to cell phones in an affected area.

We have attempted to outline a research agenda that not only examines questions about past disasters and recent technologies but also envisions what future integrated alert and warning technologies and systems might look like. As both natural and humanmade hazards occur with more frequency or severity, we hope that a future system will more readily adapt to a new set of hazards and more quickly integrate newer technologies.

> Ramesh Rao, *Chair*
> Committee on the Future of Emergency Alert
> and Warning Systems: Research Directions

Acknowledgment of Reviewers

This Consensus Study Report was reviewed in draft form by individuals chosen for their diverse perspectives and technical expertise. The purpose of this independent review is to provide candid and critical comments that will assist the National Academies of Sciences, Engineering, and Medicine in making each published report as sound as possible and to ensure that it meets the institutional standards for quality, objectivity, evidence, and responsiveness to the study charge. The review comments and draft manuscript remain confidential to protect the integrity of the deliberative process.

We wish to thank the following individuals for their review of this report:

Ellen Bass, Drexel University,
Art Botterell, California Governor's Office of Emergency Services,
Louise K. Comfort, University of Pittsburgh,
Michael Ettenberg, NAE,[1] Dolce Technologies,
W. Craig Fugate, Federal Emergency Management Agency (retired),
Dale Hatfield, University of Colorado, Boulder,
Anthony (Tony) F. Lemieux, Georgia State University,
Craig Partridge, Raytheon BBN Technologies,
Francisco Sanchez, Harris County (Texas) Office of Homeland Security and Emergency Management,

---

[1] Member, National Academy of Engineering.

Alberto Sangiovanni-Vincentelli, NAE, University of California, Berkeley, and
Sharon Wood, NAE, University of Texas, Austin.

Although the reviewers listed here provided many constructive comments and suggestions, they were not asked to endorse the conclusions or recommendations of this report nor did they see the final draft before its release. The review of this report was overseen by Phillip M. Neches, Teradata Corporation. He was responsible for making certain that an independent examination of this report was carried out in accordance with the standards of the National Academies and that all review comments were carefully considered. Responsibility for the final content rests entirely with the authoring committee and the National Academies.

Contents

	SUMMARY	1
1	UNDERSTANDING PUBLIC RESPONSE TO ALERTS AND WARNINGS Results from Earlier Decades of Research, 18 Recent Research, 28	18
2	BUILDING AN INTEGRATED ALERT AND WARNING ECOSYSTEM Need for an Integrated Alert and Warning Ecosystem, 48 Properties of an Integrated Alert and Warning System, 49 Evolution of an Integrated Alert and Warning Ecosystem, 51	45
3	A RESEARCH AGENDA Public Response to Alerts and Warnings, 56 Post-Alert Feedback and Monitoring for Emergency Organizations, 64 Technical Challenges and Their Impact, 65	56
4	CHALLENGES TO BUILDING BETTER ALERTING SYSTEMS Adoption of Alert and Warning Systems, 74 Ever Changing Technology, 76 Coupling Research with Emergency Managers and the Private Sector, 77	74

Incentives to Participate, 78
Limits in Forecasting, 78

APPENDIXES

A   Current Alert and Warning Systems and Their Characteristics   83
B   Summaries of Research Results from DHS-Supported Principal Investigators   91
C   Briefers to the Committee   127

Summary

Following a series of natural disasters, including Hurricane Katrina, that revealed shortcomings in the nation's ability to effectively alert populations at risk, Congress passed the Warning, Alert, and Response Network (WARN) Act in 2006. The law prompted the first significant changes to national alerting systems since the mid-1990s when the Emergency Alert System (EAS) replaced the Emergency Broadcast System used for radio and television alerting. The resulting Integrated Public Alert and Warning System (IPAWS) combined existing systems such as the Emergency Alert System and the National Oceanic and Atmospheric Administration Weather Radio All Hazards system and would come to include the Wireless Emergency Alerts (WEA) system,[1] which delivers short alert messages to cell phone subscribers. (See Box S.1.)

Today, new technologies such as smartphones and social media platforms offer new ways to communicate with the public, and the information ecosystem is much broader, including additional official channels, such as government social media accounts, opt-in short message service (SMS)-based alerting systems, and reverse 911 systems; less official channels, such as mainstream media outlets and weather applications on connected devices; and unofficial channels, such as first person reports via social media. Traditional media have also taken advantage of these new tools, including their own mobile applications, to extend their reach beyond broadcast radio, television, and cable; many even develop their

---

[1] This text was modified after prepublication.

> **BOX S.1**
> **Defining Alerts and Warnings**
>
> Traditionally, "alerts" have been used to indicate that something significant has happened or may happen, while "warnings" typically follow alerts and provide more detail information indicating what protective action should be taken. However, the distinction between the two terms has blurred over time, as a more nuanced understanding of people's need for information and response to that information has emerged and as new communications technologies that can transmit both types of messages have come into use. The purpose of alerts and warnings is to provide the necessary information to warn the public and effect the necessary actions that will lead to their safety and to deliver the messages to populations at risk of imminent threats with the goal of maximizing the probability that people take protective actions and minimize the delay in taking those actions.
>
> There are a variety of events that could trigger issuance of alerts and warnings, including natural hazards, such as severe weather, and humanmade events, such as terrorist attacks, active shooters, biological/chemical threats, civil unrest, and significant traffic disruptions.
>
> Alerts and warnings may be sent by government agencies, school systems (both higher education and K-12), media stations, or other information sources and sent to individuals, organizations, select groups, or broadly to the public. For example, for a hazard that affects a school, alert originators may choose to alert school administrators before alerting the community.
>
> New technologies allow alerts and warnings to be more precisely targeted to subpopulations at risk. Similarly, alerts might be sent only to those subscribing to alerts for a particular school or to everyone located in a specific geographical area. Targeting is complex; people may receive an alert even when not located in the hazard area, and people in a location may receive an alert that does not apply directly to them.
>
> Alerts and warnings may be sent before, during, or after an event. The type of information needed and the population that is at risk will shift throughout each phase of an event.

own mobile applications to deliver information. Furthermore, private companies have begun to take advantage of the large amounts of data about users they possess to detect events and provide alerts and warnings and other hazard-related information to their users; for example, Google provides alerts along with search results, and Facebook Safety Check detects emergencies and provides its users with an opportunity to register their status. Applications like Waze provide automated alerting regarding traffic situations to oncoming drivers and may also be used by government agencies to provide evacuation information. As a result, there are numerous opportunities to better deliver, target, and tailor emergency alerts.

More than 60 years of research on the public response to alerts and warnings has yielded many insights about how people respond to information that they are at risk and the circumstances under which they are most likely to take appropriate protective action. Some, but not all, of these results have been used to inform the design and operation of alert and warning systems, and new insights continue to emerge. In particular, a recent body of research, including work funded by the U.S. Department of Homeland Security (DHS; summarized in Box S.2 and described in more detail in Appendix B), has examined the implications of current and emerging technologies for the public response to alerts and warnings with a focus in part on how the public would respond to messages that could be delivered by future versions of WEA. Some of these results have already been used to enhance WEA, as evidenced by references to the research in the Federal Communication Commission's (FCC's) 2016 Report and Order on WEA,[2] including the key insight that the 90-character message length afforded by the previous WEA system was not sufficient to accommodate the quality and quantity of information necessary for yielding a quick public response.[3] Through this and other research, a good deal has been learned about how people use other tools, such as social media, during hazards and disasters.[4]

This report reviews the results of past research, considers new possibilities for realizing more effective alert and warning[5] systems, explores how a more effective national alert and warning system might be created and some of the gaps in our present knowledge, and sets forth a research agenda to advance the nation's alert and warning capabilities.

---

[2] Federal Communications Commission, 2016, Report and Order and Further Notice of Rulemaking, FCC 16-127, September 29, https://apps.fcc.gov/edocs_public/attachmatch/FCC-16-127A1.pdf.

[3] J. Sorensen and D. Mileti, 1987, Decision making uncertainties in emergency warning system organizations, *International Journal of Mass Emergencies and Disasters* 5(1):33-61.

[4] L. Palen, K.M. Anderson, G. Mark, J. Martin, D. Sicker, M. Palmer, D. Grunwalk, 2010, A Vision for Technology-Mediated Support for Public Participation and Assistance in Mass Emergencies and Disasters, Association of Computing Machinery and British Computing Society's 2010 Conference on Visions of Computer Science, Proceedings of the 2010 ACM-BCS Visions of Computer Science Conference, Article No. 8; and B.R. Lindsay, 2011, Social Media and Disasters: Current Uses, Future Options, and Policy Considerations, No. R41987, Congressional Research Service.

[5] An alert notifies the recipient that something significant has happened or may happen, and a warning, which typically follows an alert, provides more detailed information describing the event and indicates what protective action should be taken by the recipient. The distinction between alerts and warnings is not always clear-cut because a warning can also serve as an alert, and an alert may include some information about protective measures. Technology has further eroded the distinction. However, this distinction may be important as some tools are better designed to provide what has traditionally been called an alert, or vice versa.

> **BOX S.2**
> **Recent DHS-Supported Research on Alerts, Warnings, and the Wireless Emergency Alert System**
>
> This box summarizes recent research supported by the Department of Homeland Security (DHS). Appendix B also includes a lengthier description of each research project and its results.
>
> **Public Response**
>
> - *Cognitive Modeling of the Impact of Wireless Emergency Alerts* (WEAs). Experimental research on WEAs and disasters found that individuals perceive the threat of floods differently than other types of disasters.
> - *WEA Messages: Impact on Physiological, Emotional, Cognitive, and Behavioral Responses.* In a lab-based experiment, researchers assessed participants' psychophysiological, emotional, cognitive, and behavioral responses to a simulated WEA message.
> - *Results of an Integrated Approach to Geotarget At-Risk Communities and Deploy Effective Crisis Communication.* Research used ethnographic surveys and secondary data sources (public records) to examine the alert and warning needs of the diverse communities of the Mississippi Gulf Coast.
> - *Comprehensive Testing of Imminent Threat Public Messages for Mobile Devices.* The project utilized mixed methods (interviews, focus groups, and experiments) to compare first-alert WEAs to 140-character and 1,380-character messages and also tested 280-character messages.
> - *Public Response to Alerts and Warnings: Optimizing the Ability of Message Receipt by People with Disabilities.* The Center for Advanced Communications Policy (CACP) conducted research and development activities to gain a better understanding of how people with disabilities respond to WEA messages.
> - *Opportunities, Options, and Enhancements for the Wireless Emergency Alerting Service.* The primary goals of this research were to gain insight into the use of WEA by alert orginators (AOs).
>
> **Geotargeting**
>
> - *Wireless Emergency Alerts in Arbitrary Sized Target Areas: Mobile Location Aware Emergency Notification.* A new WEA geotargeting mechanism, called

AN INTEGRATED ALERT AND WARNING ECOSYSTEM FOR THE FUTURE

The development and deployment of IPAWS and WEA established a valuable new tool for public alerting, and has been credited with saving lives.[6] It leverages the ubiquity of cell phones (92 percent of American

---

[6] National Weather Service, "Wireless Emergency Alerts: Real Stories," release date May 28, 2014, https://www.weather.gov/news/130313-wea-stories.

Arbitrary-Size Location-Aware Targeting (ASLAT) was proposed, and an analysis was conducted to characterize the performance of the new mechanism and to assess the feasibility of its deployment.
- *Geo-targeting Performance of Wireless Emergency Alerts.* The objectives of this study were to evaluate the public benefit and performance trade-offs of geo-targeted WEA messages using alternative WEA antenna selection methods and to identify the optimal WEA radio frequency geo-targeted areas for imminent threat scenarios. This briefing addresses these questions for two imminent threat scenarios: tornado warnings and earthquake early warning.
- *Exploring the Effect of the Diffusion of Geo-Targeted Emergency Alerts: The Application of Agent-Based Modeling to Understanding the Spread of Messages from the WEA System.* This project took on the question of how important diffusion behavior was for understanding the value of geotargeting WEA messages.
- *Using RF Coverage to Improve Geotargeting Granularity and Accuracy for Delivery of WEA.* This project developed a geo-targeting algorithm that utilizes radio frequency (RF) cell site propagation footprints.

Technologies

- *Accessible Common Alerting Protocol Radio Data System Demonstration: Gulf Coast States.* This project demonstrated end-to-end accessible radio emergency alerting using Common Alerting Protocol (CAP) messages from FEMA's Integrated Public Alert and Warning System (IPAWS) aggregator.
- Wireless emergency alerts research from 2013 through 2016 focused on developing an integration strategy to aid AOs adopt and utilize WEA. In a follow-on project, cybersecurity risks in Commercial Mobile Service Providers (CMSPs) were assessed in relation to the possible effects on WEA and the Wireless Emergency Alerts CMSP Cybersecurity Guidelines were developed.[1]

---

[1] DHS, "Wireless Emergency Alerts (WEA) CMSP Cybersecurity Guidelines," last update July 31, 2017, https://www.dhs.gov/publication/wea-cmsp-cybersecurity-guidelines.

adults own a cell phone, and 90 percent of those owners carry their phone with them frequently).[7] However, there are many opportunities to go beyond WEA to make use of next-generation broadcast and multicast technologies, the emerging Internet of Things (IoT), and the ability of mobile devices to decide which messages to present based on user needs

---

[7] L. Rainie and K. Zickuhr, "Americans' Views on Mobile Etiquette," release date August 26, 2015, http://www.pewinternet.org/2015/08/26/americans-views-on-mobile-etiquette/.

or contextual information the device has about the user or environmental and other contextual information. Moreover, the availability of new tools and technologies will likely generate new expectations among the public. Alerts and warnings that reach people through tools and communication devices they are using and present information in a way they are accustomed to will be the most effective. For an increasingly connected population using communication media in diverse ways, any methodology that relies solely on the current (cell) broadcast technology will no longer be sufficient to serve as the primary alert and warning system.

> **FINDING:** Alert and warning systems exist within a larger communication and technical ecosystem, and government-designed and maintained systems must fit within this larger ecosystem.

> **FINDING:** A more cohesive and all-encompassing alert and warning system is needed that will integrate public and private communications mechanisms and sources of information, and continue to provide the necessary information for the purpose of preserving the health and safety of people, while being technologically agnostic—such that new technologies for alerts and warnings can be adopted quickly.

> **FINDING:** The nation's alerting capabilities, such as WEA and IPAWS, will need to evolve and progress as the capabilities of smartphones and other mobile broadband devices improve and newer technologies become available. This evolution will need to be informed by both technical research and social and behavioral science research.

EVOLUTION OF AN INTEGRATED ALERT AND WARNING ECOSYSTEM

The committee envisions an alert and warning system that continually takes advantage of new technologies and reflects new knowledge that emerges from events and research. In the near term, this may mean increasing adoption of WEA and other existing alert and warning systems, incorporation of current knowledge about public response to craft more effective alert messages, and research focusing on verifying technology implementation and may also involve adapting existing technologies—such as new technologies for delivering and geotargeting messages—for use in alert and warning systems. Long term, this will involve gaining a better understanding of existing technologies, exploring new technologies, and continued sociotechnical research to inform the design and operation of future alerting capabilities. These near- and long-term visions

SUMMARY

for an alerting system are fleshed out in this section and underpin the research agenda described in the next section.

Near Term: Adopt Existing Technologies for Alerts and Warnings

As of August 8, 2016, just under a third of U.S. counties have registered to use the Integrated Public Alert and Warning System[8] gateway, the system that allows message originators to send WEA messages. As of the same date, state or local governments had originated only 387 wireless emergency alerts since WEA came online; by comparison the National Weather Service sent approximately 2 million alerts.[9] An increased use of WEA by local emergency officials could not only mean reaching additional populations, but also increased use (for events other than weather) would improve familiarity with the systems, which could improve public response times.

Pending new FCC rules for WEA would expand the message length to 360 characters and allow the use of Web links (URLs) in messages.[10] Although DHS research studied a range of message lengths, none of the studies looked specifically at 360-character messages, the capability introduced by the FCC rulemaking, or the use of URLs, which would point users to supplemental information. As a result, although the new rules provide new opportunities for emergency managers who have struggled to provide useful information in 90 characters, research is needed to determine what information to include and how to best display additional information in a WEA message itself and on any media it links to.

WEA was developed prior to the wide use of smartphones and newer cellular network technologies. New technologies could address the shortcomings of WEA, including a host of accessibility, security, functionality, and other concerns. These advances include the following:

- *Modernize delivery technologies.* The immediate opportunity to modernize is to switch from second- or third-generation Short Message Service-based Cell Broadcast to fourth-generation long-term evolution

---

[8] IPAWS was created under Executive Order 13407 to integrate various alerting systems—Emergency Alert System, National Warning System, Wireless Emergency Alerts, and NOAA Weather Radio All Hazards—into one modern network. IPAWS takes advantage of the Common Alerting Protocol (CAP), an XML-based data format for exchanging alerts and warnings.

[9] M. Lucero, FEMA IPAWS Division, "IPAWS Evolution," presentation to the committee on August 9, 2016.

[10] See Federal Communications Commission, 2016, Report and Order and Further Notice of Rulemaking, FCC 16-127, September 29, https://apps.fcc.gov/edocs_public/attachmatch/FCC-16-127A1.pdf.

(LTE) broadcast, which provides faster delivery and longer messages, to deliver alerts.[11]

- *Diversify communications technologies in handsets* to help distribute alert messages when cellular network congestion or failure occurs. Short-range communications technology, such as Bluetooth and WiFi, could be used to forward messages locally, while FM radio provides an alternate and longer-range communication technology.

- *Support the use of location information stored in handsets to improve the precision of geotargeting.* Targeting based on relevance could be enhanced further by leveraging the ability of smartphones to determine not only where a phone is but also where it has been and thus where it is likely to be in the future.

- *Incorporate more adaptability* so that alert and warning capabilities can be upgraded more easily as understanding of public response and technology capabilities change. For example, the software on smartphones that supports receipt and presentation of WEA alerts could be moved from the operating system (which on some phones may not be frequently updated) to an app, which is more readily upgraded through the usual software update mechanisms.

- *Provide mechanisms for performance monitoring and user feedback* to facilitate studies related to perceived relevance (by seeking user feedback and/or inferring action taken), coverage (how many users did and did not receive a message), and message delivery latency.

Long Term: Build an Integrated Alert and Warning Ecosystem

In the longer term, IPAWS could be augmented so that it draws on a wide variety of data sources, enhances public understanding of emergencies and public response, and uses a wider range of potential technologies and devices for delivering messages. Envisioning such an advanced system requires exploring questions around technical feasibility and implementation and an understanding of how these tools will affect public response. However, past technical, social, and behavioral research already informs us of some of the properties that an ecosystem should have. These include the following:

- Using technologies that are privacy preserving. For example, location and other contextual information can be stored locally on a smartphone, and applications can use this information to decide when and how to display messages.

---

[11] LTE Broadcast (or multicast) provides faster delivery and supports a larger content size.

- Assuring end-to-end service availability and the validity and integrity of messages.
- Giving users as much control as possible over what kinds of messages they receive, rather than limiting them to simply on or off.
- Including metadata in alerting systems that can be used in combination with user preference to determine when and how to present alerts.
- Integrating messages across a wider array of available communication channels. For example, IPAWS messages could be made widely available as a data stream for private industry to use freely in weather applications, navigation systems, social media streams, and the like.
- Making alerting systems device agnostic and able to support more than one modality of information presentation. For example, both text and voice alerts can be provided on mobile devices.
- Reflecting a better understanding of the information needs of emergency managers to quickly analyze data generated via social media.
- Using IoT devices and other embedded sensors to detect, analyze, and categorize potential events, send alerts, and potentially automate certain protective actions.
- Incorporating available communications technologies, such as mesh networking and FM broadcast signals,[12] to increase the ability to deliver information in the event that primary communication networks fail.
- Adapting message content and format to the context and needs of the end user—for example, considering location of the device, known home location of the device owner, language of the device owner, disability status, and other context (as selected or entered by the user).

These desirable system properties and goals have the potential to inform research investments and to inform future system requirements.

A RESEARCH AGENDA

To realize the above-envisioned alert and warning system, additional research questions will need to be answered. Given that alerts and warnings are inherently interdisciplinary—both a social science phenomenon (their goal is to change public behavior) and a technical phenomenon (technology is required for their dissemination), this research agenda includes a wide range of sociotechnical questions and highlights the need for social and behavioral scientists and technologists to interact frequently with each

---

[12] Many smartphones have FM radio receiver hardware built into them. There is potential for these to be used to provide information if a cellular network is not functioning; however, enabling this function requires the consideration of a number of technical and business issues.

other. The areas of research are described below briefly and explained in greater detail in Chapter 3.

Public Response

As outlined in Chapter 1, much has been learned about the public response to alerts and warnings from years of research. However, many long-standing questions remain, and new technologies have introduced new questions. Key open topics include the following:

- *Message characteristics.* How message length and inclusion of protective guidance as hyperlinks affect public response, how to best express lead-time to a hazard, and how to best manage opt-in and opt-out preferences.
- *Accessibility.* How to most effectively provide messages in languages and dialects other than English, how to adapt to differing physical abilities, and how to account in emergency planning for disparities in access to technologies.
- *Geotargeting.* How to use the improved geotargeting capabilities afforded by WEA and the Common Alerting Protocol (CAP) best to communicate location, determine locations of interest (e.g., an individual's location might not be at risk but their residence is), make use of improving indoor location capabilities, and determine and communicate protective action based on location.
- *Community engagement.* New tools and technologies support communications among members of a community; for example, NextDoor allows people to quickly identify neighbors and communicate with those people who live nearby. NextDoor is already being used by public safety organizations to educate the public;[13] however, little is known about how such tools have been or could be used during disasters and after emergencies.

Message Characteristics

Expressing Time until Hazard Impact

Different hazards have different lead times. When too much lead-time is provided, people are less likely to follow protective guidance. Understanding how to best express lead time in WEA messages, and other alerting tools, is an important area for future research as well as the ideal lead times by hazard type.

---

[13] M. Helft, "A Facebook for Crime Fighters," *Fortune.com*, July 1, 2014, http://fortune.com/2014/07/01/nextdoor-local-neighborhood-social-network-police/.

Opt-In versus Opt-Out

Current WEA guidelines allow an individual to opt out of most emergency alerts. Past research suggests that alerts and warnings should be sent through as many channels as possible, but new research is needed to explore whether receiving the same message on numerous channels might prompt people to opt out of messages from WEA, such as third-party applications, or local text alerting systems.

Message Length and Protective Guidance in Enhanced Media Links

Although much is known about what information a message should contain, less is known about how to best communicate this information given constraints on message length and content. Research is needed, for example, to understand the public response to messages that fit into the new 360-character length, and further research is needed to determine the optimal minimum length, including personalized risk visualizations and/or URL links that can elicit the appropriate protective action from an alerted population. At this point, it is unclear what information is best included in a WEA message and what information is best included in linked content. Furthermore, concerns remain as network congestion could be caused by people accessing an included link within seconds of receiving a WEA that includes a URL.[14]

Accessibility

Language and Dialect

What technical challenges exist in transmitting messages in multiple languages or relying on the receiving device to translate messages? Are there practical limits to the number of different languages that can be supported? Additionally, key language elements, such as descriptions of protective actions, may be challenging to translate accurately to various languages and dialects. Research is needed to understand whether templates can be created so that messages can be automatically translated with sufficient fidelity.

---

[14] Federal Communications Commission, "Improving Wireless Emergency Alerts and Community-Initiated Alerting," release date November 19, 2015, https://apps.fcc.gov/edocs_public/attachmatch/FCC-15-154A1.pdf.

Adapting to Differing Abilities

A variety of technologies, such as vibration cadences and Braille interfaces, have been developed to allow mobile phones to be used by a wide range of physically and cognitively challenged message recipients. Research is needed to understand how best to customize the content and means of delivery to physically and cognitively challenged individuals. What other technologies exist to support information dissemination to differently abled individuals? How can protective action instructions be tailored or customized to support diverse populations—including those of differing ages and abilities—and their caregivers?

Digital Divide

Although a large and growing portion of the population uses smartphones, there are still others who cannot afford or choose not to use them. Considering the diversity in communication habits and availability of technology, alert and warning systems will need to consider various technologies to reach individuals facing hazards.

Geotargeting

Communicating Location

What graphics most effectively indicate that an individual is in an at-risk location? How can visualizations be used to best illustrate the location of the message receiver relative to the area of impact? What is the best way to communicate to someone who is unfamiliar with the area they are in?

Determining Locations of Interest

Individuals want to be alerted not only when they are at risk, but also, for example, when their children may be at risk or their home may be at risk. How can locations of interest be determined and updated automatically rather than manually specified and updated by the end-user?

Location-Based Protective Action

The best protective action—for example, shelter in place versus evacuate—for an individual may vary across the affected area. Furthermore, individuals could be assigned diverse routes to enhance evacuation traffic flows. What are the technical challenges to providing such precision?

What are the limitations to implementing these for disaster response? How might we encourage use of these tools?

In-Building Location

Knowledge of a person's location within a building could be used to determine the best evacuation route or if the individual should instead shelter in place. Limited indoor location capabilities are already being deployed in some areas, chiefly for marketing purposes; determining building floor is a bit harder. What location techniques are emerging, and how could they best be used?

Hazard and Alerting Education

Very limited research has been performed looking at what makes hazard and alerting public education effective. Research to date has found current public education campaigns are generally ineffective because they are not specific enough and do not contain content that motivates behavior change.[15] More research is needed to determine how to motivate behavior change as well as what other factors contribute to successful public disaster education campaigns.

Post-Alert Feedback and Monitoring

Technology is needed that solicits feedback from message recipients to help understand better who has received alert messages, how the public is responding to the messages, and what additional information might be needed. WEA's cell broadcast technology may not reach all phones in the target area, and today's systems are one-way, meaning that alert originators have no way to know who has actually received an alert. An acknowledgment mechanism would be a useful element as part of any feedback and monitoring system. Some information about the public response can also be obtained using tools that extract information from social media. More direct feedback mechanisms could be built into alerting applications on mobile devices, and these tools will need to be more readily available. Perhaps more importantly, research is needed to under-

---

[15] B.J. Adame, and C.H. Miller, 2015, Vested interest, disaster preparedness, and strategic campaign message design, *Health Communication* 30(3):271-281; J.D. Fraustino, and L. Ma, 2015, CDC's use of social media and humor in a risk campaign – Preparedness 101: Zombie apocalypse, *Journal of Applied Communication Research* 43(2):222-241; and M.M. Turner, and J.C. Underhill, 2012, Motivating emergency preparedness behaviors: The differential effects of guild appeals and actually anticipating guilty feelings, *Communication Quarterly* 60(4):545-559.

stand what information would be more helpful to emergency managers. Tools, including those that employ machine learning and other artificial intelligence techniques, are also needed to quickly understand and process feedback to ensure emergency managers are not overwhelmed with information.

A future alerts and warnings ecosystem that includes consistent, well-understood, and insightful measurements could inform (and improve) response to future hazards. Such a data-driven experimental framework would be of great interest to multiple stakeholders, including emergency managers and researchers. By building measurement into the alerts and warning system itself, researchers could gain supporting evidence for findings made in lab studies (e.g., which message length is appropriate? should we include a map or not?). Feedback during the life cycle of a hazard could also be integrated into future responses within the same incident. For example, low response rates to an initial message could lead to more aggressive message content in a follow-on message.

Alerting System Trustworthiness

Existing and new technologies can increase the likelihood that messages reach populations at risk, enable richer interactions between emergency managers and those at risk, and automate certain time-sensitive functions. At the same time, current and future systems will be at risk of malicious attacks and attention will be needed to protect individual privacy.

Delivery Technologies

Today, WEA is designed to use only cellular communications. While there are still several technical research questions around cell broadcast technologies, such as the use of next-generation networks, cell phones can also receive data through a variety of other wireless communications technologies that could be adapted for message dissemination. Additionally, during hazards, some cellular networks may not function properly, so other technologies are needed to deliver messages (e.g., peer-to-peer, FM radio). Battery life management on end-user devices is also a rich area of research.

Role of Connected Devices

As the IoT grows, more devices in homes and throughout the environment will be available not only as an alerting channel but also to detect emergencies and potential risk. It may be helpful to ensure that many connected devices that have a sensory output can be triggered to provide an alert or warning. To make the most effective use of these opportunities, several questions around aggregating data, potential use of automation, which devices are best suited for alerting, and the potential role of IoT devices in milling. Machine learning and other artificial intelligence techniques will play a role in the ability to automate the sending of alerts in short-fuse events, such as earthquakes and active shooter situations, and also to provide responders with better information during and post-events.

Trust, Security, and Privacy

A system that instructs large populations to take a particular action may represent a significant target for attacks on service availability, compromises of the integrity of valid messages, and spoofed messages.[16] As emergency managers begin harnessing information—including personal and geographically relevant information—from social media, security and privacy concerns will increase. How can we take advantage of these tools while still protecting end-user privacy?

Furthermore, in a system that makes use of user-generated (public-generated) content, misinformation becomes an increasing concern as well. Quickly detecting and correcting poor information will be valuable system capabilities.

CHALLENGES TO BUILDING BETTER ALERTING SYSTEMS

Beyond the specific research topics listed above, the committee noted several challenges to building better alert and warning systems.

Slow Adoption of New Systems

Reasons for lack of adoption include system costs for jurisdictions and message originators' education, but even those with access to the IPAWS gateway can be hesitant to use the system. Moreover, in smaller jurisdictions, sending alerts may be a part-time job, and a person may only be active in the emergency response community during events; in the largest jurisdictions, public alerting may be the responsibility of a large team of individuals who are trained emergency management professionals immersed in disaster response full time.

---

[16] C. Woody, Software Engineering Institute of Carnegie Mellon University, "SEI Wireless Emergency Alerts (WEA) Research 2013 through 2016," presentation to the committee on September 1, 2016.

Limitations on Weather Forecasts and Other Information about Natural Hazards

Agencies that distribute weather-related messages at the state, local, regional, or federal level must ultimately rely on forecasts and weather information from the National Weather Service and National Oceanic and Atmospheric Administration and information provided by the U.S. Geological Survey. These agencies in turn rely on the infrastructure that collects, models, and distributes information about weather, earthquakes, air quality, and other environmental conditions. Information provided by these agencies also supports an array of private-sector alerting services. Effective alerting depends on modeling and data collection and analysis capabilities being maintained and advanced.

Ever-Changing Technology

The array of information and communications technologies used by the public are continually evolving. At the same time, old and new technologies coexist for long periods of time. To reach the majority of individuals, alerting systems must not only evolve but continue to make use of legacy technologies. Furthermore, both the technology of emergency alerts and citizens' capacity to comprehend the alerts and use messaging functions also continue to evolve. The interaction between the developing technologies and citizens' capacity to use these technologies effectively on a community scale is itself an issue for future research.

Difficulty of Interdisciplinary Research and Converting Research to Practice

Public response to alerts is a highly interdisciplinary activity, and it is also an activity closely coupled to the practice of emergency management, which takes place primarily at the state and local level in the United States. Yet technologists, social science researchers, and emergency managers have had few opportunities for ongoing interactions to consider how to apply current knowledge or fill gaps in our understanding.

Incentives to Participate

An alert and warning ecosystem incorporates numerous official sources of information as well as numerous other information providers, such as social media companies, navigation companies, local media, and hardware makers. For example, WEA relies on cellular service providers to implement the necessary capabilities in their infrastructure and for cell phone manufacturers to include the necessary software in smartphones (although participation is voluntary, all major carriers currently participate). Incorporating these various pieces, and ensuring that information about how the system is working is shared, will be an increasing challenge. How do we encourage openness among stakeholders and encourage participation by those who operate other valuable computer and communications capabilities?

***

Our nation's ability to respond effectively to natural hazards and humanmade disasters depends on our ability to deploy improved alerting systems that take advantage of new technologies, informed by a better understanding of the way in which the public uses and responds to these systems. Doing so will depend on addressing the challenges and research areas listed above.

1

Understanding Public Response to Alerts and Warnings

This chapter summarizes research on public response to alerts and warnings. It starts by reviewing results of research from the 1970s to the 1990s and then turns to more recent research, including work sponsored by the Department of Homeland Security (DHS) that explored public response in the context of Wireless Emergency Alerts (WEA) system and other emerging technologies.[1] The research summarized in this chapter is only a subset of a large body of work done on emergency alerts and warnings. Summarized research was selected based on relevance, with an emphasis on work funded by DHS (in response to the committee's charge). Attempts were made to summarize not only long-standing research but also research that looks at newer tools and technologies, such as mobile devices, WEA, and social media.

RESULTS FROM EARLIER DECADES OF RESEARCH

There is almost always a delay between when an alert is received and when the recipient takes a protective action. That time gap is known as the protective action initiation (PAI) time (Box 1.1 outlines the warning process). One major area of research aims to shorten that time period. This section starts by discussing factors that influence a person's PAI time, including milling, reunification with intimates, and the time it takes to

---

[1] Principal investigators funded by the Department of Homeland Security Science and Technology directorate were asked to provide brief summaries of their work to the committee. These summaries are printed in full in Appendix B.

> **BOX 1.1**
> **The Warning Process for Imminent Events**
>
> Public response to an alert or warning message is only one part of an emergency alert and warning cycle. Below is a simplified cycle of what occurs during an alert and warning exercise:
>
> 1. The event must be detected.
> 2. A decision on whether or not to warn the public must be made.
> 3. The public must receive and subsequently understand the warning.
> 4. The public must have been given options of actions to take or safe places to go.
> 5. The public must choose to take action.
>
> Imminent threats include both natural and humanmade disasters, like severe weather conditions, terrorist attacks, active shooters, or biological/chemical threats.

prepare for the possible protective action. It then turns to how message content, message context, and message receiver characteristics can also impact PAI.

The key PAI question is "What delays people from taking a protective action upon receipt of a first alert/warning or observation of environmental or social cues?" Between the point of receiving a message and the point of taking a protective action, people generally engage in a variety of activities that reconstructs their perception of safety into a reception of personal risk, creating the delay between the alert and the action.

Milling, Reunification, and Preparedness

People often seek confirmation from others regarding alerts and warnings, which is a process referred to as milling. Through milling, people form ideas concerning personal safety, risk, and what to do about it. Individuals during this time engage in a series of activities designed to increase their comprehension of the event, which includes understanding, believing, personalizing, deciding, and searching and confirming (Box 1.2). Milling occurs regardless of the hazard type, the warning delivery technology used, or the source of the warning. Hence basic human nature creates a response gap for most people between getting an initial alert/warning and initiating a protective action.

In addition to milling, the response gap is also affected by two additional factors: reunification of intimates and protective action preparation.

> **BOX 1.2**
> **The Process of Milling**
>
> Milling is the process of seeking information from others regarding alert and warning messages. Individuals milling go through the following steps upon receiving an alert or warning:
>
> - Understanding[1]—This type of understanding refers not to the correct interpretation of what is heard, but to the personal attachment of meaning to the message. The meaning of a message can vary between different people.
> - Believing[2]—People determine whether or not to believe that a warning is real.
> - Personalizing[3]—Even if a person understands and believes a warning, they will not act if they do not believe that they, their families, or their group are targets of the warning (also referred to as risk personalization).
> - Deciding[4]—People will decide what (if anything) to do about the risk. Deciding is recognized as part of the pre-response sense-making process, but it has rarely been explicitly studied as a variable in and of itself.
> - Searching and confirming[5]—Searching for additional and confirming information is a basic post-warning receipt but pre-protective action-taking behavior.
>
> ---
>
> [1] Studies that document the effect of non-message factors on understanding include J.C. Diggory, 1956, Some consequences of proximity to a disease threat, *Sociometry* 19:47-53; J. Nehnevajsa, 1985, *Western Pennsylvania: Some Issues in Warning the Population Under Emergency Conditions*, University of Pittsburgh, Center for Social and Urban Research.
> Studies that document the effect of understanding on protective action-taking behavior include G. Hammarstrom-Tornstam, 1977, Varingprocessen (Warning process), *Disaster Studies 5*, Uppsala, Sweden: University of Uppsala; and R.W. Perry, 1982, *The Social Psychology of Civil Defense*, Lexington, MA: Lexington Books.
> [2] Studies that document the effect of message content and style factors on belief include E.J. Baker, 1979, Predicting response to hurricane warnings: A reanalysis of data from four studies, *Mass Emergencies* 4:9-24; G. Rogers, 1985, *Human Components of Emergency Warning*, Pittsburgh, PA: Center for Social and Urban Research, University of Pittsburgh; and University of Oklahoma Research Institute, 1953, *The Kansas City Flood and Fire of 1951*, Baltimore, MD: Operations Research Office, Johns Hopkins University.
> Studies that document the effect of non-message factors on believing include J. Nigg, 1987, Communication and behavior: Organizational and individual response to warnings, pp. 103-

People are less likely to initiate protective action until all members of their immediate family have reunified.[2] This normally involves waiting

---

[2] T. Drabek and J. Stephenson III, 1971, When disaster strikes, *Journal of Applied Social Psychology* 1(2):187-203; R. Mack and G. Baker, 1961, *The Occasion Instant: The Structure of Social Responses to Repeated Air Raid Warnings*, Disaster Study No. 15, Washington, DC: National Research Council, National Academy of Sciences; and R. Perry, 1987, Disaster Preparedness

117 in *Sociology of Disasters: Contributions of Sociology to Disaster Research* (R.R. Dynes, B. DeMarchi, and C. Pelanda, eds.), Milano, Italy: Franco Angeli; J. Ponting, 1974, It can happen here: A pedagogical look at community coordination to response to a toxic gas leak, *Emergency Planning Digest* 1:8-13.

Studies that document the effect of belief on protective action-taking behavior include E. Danzig, P. Thayer, and L. Galater, 1958, *The Effects of a Threatening Rumor on a Disaster-Stricken Community*, Disaster Study No. 10. Washington, DC: Disaster Research Group, National Academy of Sciences; T. Hodler, 1982, Residents preparedness and response to the Kalamazoo tornado, *Disasters* 6(1):44-49.

[3] Studies that document the effect of message content and style factors on personalizing include R. Perry, 1979, Evacuation decision-making in natural disasters, *Mass Emergencies* 4:25-38; and R. Perry, M. Lindell, and M. Greene, 1981, *Evacuation Planning in Emergency Management*, Lexington, MA: Lexington Books.

Studies that document the effect of non-message factors on personalizing include C. Flynn, 1979, *Three Mile Island Telephone Survey: Preliminary Report on Procedures and Findings*, Tempe, AZ: Mountain West Research; R. Hansson, D. Noulles, and S. Bellovich, 1982, Knowledge, warning, and stress, *Environment and Behavior* 14(2):171-185.

Studies that document the effect of personalizing on protective action-taking behavior include R. Perry, M. Lindell, and M. Greene, 1981, *Evacuation Planning in Emergency Management*, Lexington, MA: Lexington Books; B. Phillips and B. Morrow, 2007, Social science research needs: Focus on vulnerable populations, forecasting, and warnings, *Natural Hazards Review* 8(3):61-68; T. Sellnow, D. Sellnow, D. Lane, and R. Littlefield, 2012, The value of instructional communication in crisis situations: Restoring order to chaos, *Risk Analysis* 32:633-634.

[4] Deciding is recognized as part of the pre-response sense-making process, but it has rarely been explicitly studied as a variable in and of itself (M. Lindell and R. Perry, 2012, The Protective Action Decision Model: Theoretical modifications and additional evidence, *Risk Analysis* 32(4):616-632; J. Sorensen, and D. Mileti, 1987, Decision making uncertainties in emergency warning system organizations, *International Journal of Mass Emergencies and Disasters* 5(1):33-61. Studies on alternative protective action decision include B. Vogt, and J. Sorensen, 1999, *Description of Survey Data Regarding the Chemical Repackaging Plant Accident, West Helena, Arkansas*, ORNL/TM-13722, Oak Ridge, TN: Oak Ridge National Laboratory; M. Dombroski, B. Fischhoff, and P. Fischbeck, 2006, Predicting emergency evacuation and sheltering behavior: A structured analytical approach, *Risk Analysis* 26(6):1675-1688.

More complex models of evacuation compliance have been developed, with risk perception as the central focus and with more reliable indicators of evacuation behavior (M. Lindell and R. Perry, 2012, The Protective Action Decision Model: Theoretical modifications and additional evidence, *Risk Analysis* 32(4):616-632; and K. Dow, and S. Cutter, 1998, Crying wolf: Repeat responses to hurricane evacuation orders, *Coastal Management* 26(4):237–252.

[5] M. Lindell and R. Perry, 2012, The Protective Action Decision Model: Theoretical modifications and additional evidence, *Risk Analysis* 32(4):616-632; and J. Sorensen, and D. Mileti, 1987, Decision making uncertainties in emergency warning system organizations, *International Journal of Mass Emergencies and Disasters* 5(1):33-61.

for family members to assemble, going home from work, or picking up children from schools, day care, or other locations. Both early and recent research shows that households with children or pets are inhibited from

---

and Response Among Minority Citizens, pp. 135-151 in *Sociology of Disasters: Contributions of Sociology to Disaster Research* (R.R. Dynes, B. DeMarchi, and C. Pelanda, eds.), Milano, Italy: Franco Angeli.

taking a suggested protective action until they are reunited with children[3] and are permitted to bring their pets.[4]

There is also the protective action preparation, which is the time required to organize resources to implement the protective action. This period can involve a variety of actions, depending on the hazard. Actions can range from assembling resources such as emergency supplies, clothing, food and water; filling the evacuation vehicle with gas; or securing the home from hazard impacts or human intrusion. It is logical to hypothesize that people who are better prepared (e.g., they keep shelter kits in cars or have emergency food and water supplies pre-packed) will be less likely to delay taking the recommended protective action. Similarly, the more a message recipient knows about the hazard, the protective actions associated with that hazard, or how warnings about that hazard might be delivered to them in their specific location, the more likely they are to act promptly.[5] Members of groups that are socially isolated are more likely to either fail to respond or not respond promptly to a message.

Message Characteristics Influencing Protective Action Initiation Times

Decades of work has identified that a variety of message characteristics—including content, style, length, delivery, and type of recommended protective action—influence public response.

---

[3] T. Carter, S. Kendall, and J. Clark, 1983, Household response to warnings, *International Journal of Mass Emergencies and Disasters* 9(1): 94-104.; R. Lachman, M. Tatsuoka, and W. Bonk, 1961, Human behavior during the tsunami of May, 1960, *Science* 133:1405-1409; K. Wilkinson and P. Ross, 1970, *Citizens Response to Warnings or Hurricane Camille*, Report No. 35. State College: Social Science Research Center, Mississippi State University.

[4] T.E. Drabek and K. Boggs, 1968, Families in disaster: Reactions and relatives, *Journal of Marriage and the Family* 30:443-451; S. Heath and M. Champion, 1996, Human health concerns from pet ownership after a tornado, *Prehospital and Disaster Medicine* 11(1):67-70; S. Heath, P. Kass, A. Beck, and L. Glickman, 2001, Human and pet-related risk factors for household evacuation failure during a natural disaster, *American Journal of Epidemiology* 153(7):659-665; A. Edmonds and S. Cutter, 2008, Planning for pet evacuations during disasters, *Journal of Homeland Security and Emergency Management* 5(1); L.K. Zottarelli, 2010, Broken bond: An exploration of human factors associated with companion animal loss during Hurricane Katrina, *Sociological Forum* 25(1):110-122; T. Litman, 2006, Lessons from Katrina and Rita: What major disasters can teach transportation planners, *Journal of Transportation Engineering* 132(1); R.J. Blendon, J.M. Benson, C.M. DesRoches, K. Lyon-Daniel, E.W. Mitchell, and W.E. Pollard, 2007, The public's preparedness for hurricanes in four affected regions, *Public Health Reports* 122:167-176.

[5] D. Glik, K. Harrison, M. Davoudi, and D. Riopelle, 2004, Public perceptions and risk communication for botulism, *Biosecurity and Bioterrorism: Biodefense Strategy, Practice, and Science* 2(3):216-223; J. Haas, H. Cochrane, and D. Eddy,1977, Consequences of a cyclone on a small city, *Ekistics* 44(260):45-50; and M. Lehto and J. Miller, 1986, *Warnings, Vol. I: Fundamentals, Design, and Evaluation Methodologies*, Ann Arbor, MI: Fuller Technical Publications.

Message Content

The research record provides repetitive evidence that public warning messages are more likely to motivate appropriate and timely public protective actions if the warning messages contain information on five topics: guidance, time, location, hazard and consequences, and source.[6]

Research has shown that people increase responsiveness when they receive guidance on exactly what to do to maximize their safety and how to do it, including information on what factors to consider when deciding on whether or not to stay.[7] Time and location are important content for a message because it informs people when they should begin taking a protective action and by when they should have it completed,[8] in addition to saying exactly who should and who should not take action in terms that the public can readily understand (e.g., physical geographical boundaries of those who will be affected).[9] Messages also contain information about the specific hazard the warning is for and the potential consequences (e.g., hurricanes can cause flooding), which gives context to the guidance being given in message. To take that guidance, however, people must trust the source. Individuals will turn to their neighbors and family to verify warning messages.

Message Style

Research from the 1990s provides strong evidence that public warning messages with certain style elements work best. Those factors are typically referred to as warning or message style and include clarity, specificity, accuracy, certainty, and consistency. People are more receptive to messages that are free of jargon and are written in words that most people can understand.[10] People also want specific language that gives

---

[6] D. Mileti and J. Sorensen, 1990, *Communication of Emergency Public Warnings: A Social Science Perspective and State-of-the-Art Assessment*, Oak Ridge, TN: Oak Ridge National Laboratory, U.S. Department of Energy.

[7] T. Drabek, 1999, Understanding disaster warning responses. *Social Science Journal* 36(3):515-523; D. Mileti and C. Fitzpatrick, 1991, The causal sequence of risk communication in the Parkfield Earthquake Prediction Experiment, *Risk Analysis* 12(3):393-399; J. Sorensen, 1991, When shall we leave: Factors affecting the timing of evacuation departures, *International Journal of Mass Emergencies and Disasters* 9(2):153-165.

[8] Ibid.

[9] T. Drabek, 1999, Understanding disaster warning responses, *Social Science Journal* 36(3):515-523; and D. Mileti and C. Fitzpatrick, 1991, The causal sequence of risk communication in the Parkfield Earthquake Prediction Experiment, *Risk Analysis* 12(3):393-399.

[10] L. Bellamy and P.I. Harrison, 1988, An Evacuation Model for Major Accidents, Paper presented at the IBC Conference on Disaster and Emergencies, London, April; J. Nigg, 1987, Communication and Behavior: Organizational and Individual Response to Warnings, pp. 103-117 in *Sociology of Disasters: Contributions of Sociology to Disaster Research* (R.R. Dynes,

precise and non-ambiguous information about the area(s) at risk, how much time they have to engage in protective actions before impact, and the source of the message.[11] Timely and accurate information that is complete and free from errors becomes important during these events. People may disregard a message or consider the source(s) to be non-credible if individuals come to learn or suspect that they are not receiving the truth in its entirety.[12] Transparency and honesty regarding a hazard enhances the perception of accuracy.[13] As noted in a study published in *Information Communication & Society*, facts relating to the hazard need to be stated "authoritatively, confidently, and with certainty, even in circumstances in which there is ambiguity about message content factors and especially about the protective action the public is being asked to take."[14] These messages also explain that, even though physical details about the hazard are changing, experts agree on the protective actions people should take.[15] Messages also need to be externally consistent, for example, by explaining any changes that may have occurred since the previous message, as well as be internally consistent.[16]

B. DeMarchi, and C. Pelanda, eds.), Milano, Italy: Franco Angeli; D. Mileti and E. Beck, 1975, Communication in crisis: Explaining evacuation symbolically, *Communication Research* 2:24-29; and B. McLuckie, 1975, *Warning: A Call to Action*, Washington, DC: U.S. Weather Service.

[11] M. Lindell and R. Perry, 1987, Warning mechanisms in emergency response systems, *International Journal of Mass Emergencies and Disasters* 5(2):137-153; G. Rogers, 1985, *Human Components of Emergency Warning*, Pittsburgh, PA: Center for Social and Urban Research, University of Pittsburgh; P. Houts, M. Lindell, T. Weittu, P. Clearly, G. Tokuhata, and C. Flynn, 1984, The Protective Action Decision Model applied to evacuation during the Three-Mile Island crisis, *International Journal of Mass Emergency and Disasters* 2(1):27-39; and R. Perry, M. Lindell, and M. Greene, 1981, *Evacuation Planning in Emergency*, Lexington, MA: Lexington Books.

[12] D. Mileti, 2012, Chapter 26, Public Response to Flood Warnings, in *Coping with Floods* (G. Rossi, N.B. Harmanciogammalu, and V. Yevjevich, eds.), NATO ASI Series, Series E: Applied Sciences, Vol. 257. Dordrecht, The Netherlands: Kluwer Academic Publishers.

[13] D. Mileti, T. Drabek, and J. Haas, 1975, *Human Systems in Extreme Environments: A Sociological Perspective*, Boulder, CO: Institute of Behavioral Science, University of Colorado.

[14] J. Sutton, E.S. Spiro, B. Johnson, S. Fitzhugh, B. Gibson, and C.T. Butts, 2014, Warning tweets: Serial transmission of messages during the warning phase of a disaster event, *Information, Communication & Society* 17:6, 765-787, doi:10.1080/1369118X.2013.862561.

[15] D. Mileti and P. O'Brien, 1992, Warnings during disaster: Normalizing communicated risk, *Social Problems* 39(1):40-57; R. Perry, M. Lindell, and M. Greene, 1982, Crisis communications: Ethnic differentials in interpreting and acting on disaster warnings, *Social Behavior and Personality* 10(1):97-104; and R. Turner, J. Nigg, D. Paz, and B. Young, 1979, *Earthquake Threat: The Human Response in Southern California*, Los Angeles, CA: Institute for Social Science Research, University of California.

[16] J. Nigg, 1987, Communication and behavior: Organizational and individual response to warnings, pp. 103-117 in *Sociology of Disasters: Contributions of Sociology to Disaster Research* (R.R. Dynes, B. DeMarchi, and C. Pelanda, eds.), Milano, Italy: Franco Angeli.; A. Chiu, L.E Escalante, J.K. Mitchell, D.C. Perry, and T.A. Schroeder, 1983, *Hurricane Iwa, Hawaii, November 23, 1982*, Washington, DC: National Academy of Sciences; R. Perry, 1983, Population evacu-

Message Length

Recent research indicates that the length of a message plays a critical role in influencing people's understanding, belief, decision making, risk personalization, and the amount of time they delay initiating a protective action.[17]

Delivery Method

Channels of delivery can be viewed by the public as official,[18] credible,[19] or familiar,[20] or involve human interaction[21] and are each effective in some settings, but not all.

Hazard Type

Individuals may have preconceived ideas about particular hazards that affect the type of message they are willing to receive and forward

---

ation in volcanic eruptions, floods, and nuclear power plant accidents: Some elementary comparisons, *Journal of Community Psychology* 11(1):36-47; and C. Flynn, 1979, *Three Mile Island Telephone Survey: Preliminary Report on Procedures and Findings*, Tempe, AZ: Mountain West Research.

[17] D. Glik, K. Harrison, M. Davoudi, and D. Riopelle, 2004, Public perceptions and risk communication for botulism, *Biosecurity and Bioterrorism: Biodefense Strategy, Practice, and Science* 2(3):216-223; H. Bean, B.F. Liu, S. Madden, J. Sutton, M.M. Wood, and D.S. Mileti, 2016, Disaster warnings in your pocket: How audiences interpret mobile alerts for an unfamiliar hazard, *Journal of Contingencies and Crisis Management* 24(3): 136-147.

[18] R. Perry and M. Greene,1982, The role of ethnicity in the emergency decision-making process, *Sociological Inquiry* 52(Fall):309-34; G. Rogers, 1985, *Human Components of Emergency Warning*, Pittsburg, PA: Center for Social and Urban Research, University of Pittsburg; T. Saarinen and J. Sell, 1985, *Warning and Response to the Mount St. Helens Eruption*, Albany, NY: State University of New York Press.

[19] S. Cutter, 1987, Airborne toxic releases: Are communities prepared? *Environment* 29(6):12-17, 28-31; R. Perry, 1987, Disaster Preparedness and Response Among Minority Citizens, pp. 135-151 in *Sociology of Disasters: Contributions of Sociology to Disaster Research* (R.R. Dynes, B. DeMarchi, and C. Pelanda, eds.), Milano, Italy: Franco Angeli.; R. Stallings,1984, Evacuation behavior at Three Mile Island, *International Journal of Mass Emergencies and Disasters* 2:11-26.

[20] M. Lindell and R. Perry, 1987, Warning mechanisms in emergency response systems, *International Journal of Mass Emergencies and Disasters* 5(2):137-153; R. Perry and M. Lindell, 1986, *Twentieth-century Volcanicity at Mt. St. Hellens: The Routinization of Life Near and Active Volcano*, Tempe, AZ: School of Public Affairs, Arizona State University; R. Perry and M. Greene,1982, The role of ethnicity in the emergency decision-making process, *Sociological Inquiry* 52(Fall):309-34.

[21] S. Cutter, 1987, Airborne toxic releases: Are communities prepared? *Environment* 29(6):12-17, 28-31; J. Gray, 1981, Characteristic patterns of and variations in community response to acute chemical emergencies, *Journal of Hazardous Materials* 4:357-365; R. Perry, M. Lindell, and M. Greene, 1981, *Evacuation Planning in Emergency Management*, Lexington, MA: Lexington Books.

to their networks. A study took subjects and had them look at a series of WEA and Twitter[22] messages regarding five different types of disasters (flood, blizzard, hurricane, gas leak, and tornado). In the study, participants were more predisposed to share WEA messages or disaster tweets on Twitter expressing a dismissive sentiment about floods more than the other types of hazards, although overall, all subjects were highly responsive to the disaster messages and shared them a majority of the time.[23] The study also found that subjects have different responses to different hazard types based on their perceived amount of danger or damage associated with that disaster, for example, on a psychological level,[24] subjects perceive the threat posed by a flash flood differently than the other hazards in the study both while reading alerts about floods and when they were about to watch a video concerning floods.[25]

Context Characteristics Influencing Protective Action Initiation Times

Research shows that there are environmental and social cues that influence how people interpret alerts and warnings, which in turn, influences their PAI times. Environmental cues are indicators in one's environment that reinforce the presence of the hazard. People are more likely to conclude the need to take a protective action stated in an alert or warning message if environmental cues are present that reinforce the presence of the hazard.[26] Similarly, social cues are indicators in one's social environ-

---

[22] Twitter has recently increased the character limit for tweets from 140 characters to 280 characters for all users, with the exception of those tweeting in Chinese, Japanese, and Korean languages (A. Heath, 2017, "Twitter is turning on longer 280-character tweets for everyone," *Business Insider*, http://www.businessinsider.com/twitter-280-character-tweets-for-everyone-2017-11).

[23] C.D. Corley, Pacific Northwest National Laboratory, "Cognitive Modeling of the Impact of Wireless Emergency Alerts," presentation to the committee on September 1, 2016.

[24] The study collected 20-channel electroencephalography data in order to evaluate perception and responses.

[25] C.D. Corley, Pacific Northwest National Laboratory, "Cognitive Modeling of the Impact of Wireless Emergency Alerts," presentation to the committee on September 1, 2016.

[26] J. Averill, D. Mileti, R. Peacock, E. Kuligowski, N. Groner, G. Proulx, and H. Nelson, 2005, *Predicting Evacuation Delay in the World Trade Center: Occupant Behavior, Egress, and Emergency Communications: Federal Building and Fire Safety Investigation of the World Trade Center Disaster*, NIST NCSTAR 1-7. Washington, DC: U.S. Department of Commerce, National Institute of Standards and Technology; C. Flynn, 1979, *Three Mile Island Telephone Survey: Preliminary Report on Procedures and Findings*, Tempe, AZ: Mountain West Research; M. Lindell and R. Perry, 2012, The Protective Action Decision Model: Theoretical modifications and additional evidence, *Risk Analysis* 32(4):616-632; R. Mack and G. Baker, 1961, *The Occasion Instant: The Structure of Social Responses to Repeated Air Raid Warnings*, Disaster Study No. 15, Washington, DC: National Research Council, National Academy of Sciences; G. Rogers and J. Nehnevajsa, 1987, Warning human populations of technological hazards, pp. 357-362 in

ment that reinforce the presence of the hazard and the need to take the protective action recommended in an alert or warning. People are more likely to initiate a protective action when social cues are present.[27] These cues include seeing and hearing about others who are taking the recommended protective action.

How much time people have before a hazard strikes and the expected impact intensity also play a role in PAI times. Research has found that the amount of time people have to initiate an action before an event strongly influences PAI.[28] As that time decreases, so does the PAI time. In hurricanes, people are quicker in taking protective actions as the impact time draws closer.[29] Additionally, in hurricanes, more people decide to evacuate when strong storms are forecasted (impact intensity). People respond faster to hazards that are "dreaded" than ones that are "known." For example, chemical accidents and nuclear power elicit a rapid response to avoid contamination.[30] Hazards that are perceived as posing a serious threat elicit faster response.[31]

Research has also found that PAI times may be affected by the time of day and an individual's location and activity when a message is received. However, there are currently no empirical studies on community-wide alert and warning events that establish or dismiss that initiation of a protective action, for example, evacuation, would take more time during the nighttime compared to daytime. Nor is there appropriate documentation to identify how a person's location and activity impact protective initiation time. This includes if a person is sleeping, working, shopping, traveling, or engaged in recreation.[32]

ANS Topical Meeting on Radiological Accidents: Perspectives and Emergency Planning, Oak Ridge, TN: Oak Ridge National Laboratory.

[27] S. Cutter, 1987, Airborne toxic releases: Are communities prepared? *Environment* 29(6):12-17, 28-31; and R. Dynes and E. Quarantelli, 1976, The family and community context of individual reactions to disaster, pp. 231-245 in *Emergency and Disaster Management: A Mental Health Sourcebook* (H. Parad, H. Resnik, and L. Parad, eds.), Bowie, MD: Charles Press.

[28] M. Lindell and R. Perry, 1992, *Behavioral Foundations of Community Emergency Planning*, Washington DC: Hemisphere Press.

[29] E.J. Baker, 1987, *Warning and Evacuation in Hurricanes Elena and Kate*, Tallahassee, FL: Department of Geography, Florida State University.

[30] M. Lindell and R. Perry, 1992, *Behavioral Foundations of Community Emergency Planning*, Washington DC: Hemisphere Press.

[31] D. Sorensen and D. Mileti, 1987, Decision making uncertainties in emergency warning system organizations, *International Journal of Mass Emergencies and Disasters* 5(1):33-61.

[32] E. Baker, 1979, Predicting response to hurricane warnings: A reanalysis of data from four studies, *Mass Emergencies* 4:9-24; R. Clifford, 1956, *The Rio Grande Flood: A Comparative Study of Border Communities*, Disaster Study No. 17, Washington, DC: National Research Council, National Academy of Sciences; P. Houts, M. Lindell, T. Weittu, P. Clearly, G. Tokuhata, and C. Flynn, 1984, The Protective Action Decision Model applied to evacuation during the Three-Mile Island crisis, *International Journal of Mass Emergency and Disasters* 2(1):27-39;

Message Receiver Characteristics Influencing Protective Action Initiation Times

Over time, research shed more light on the human factors—such as age, gender, ethnicity, race, disabilities, and socioeconomic status—that influence response to warning messages. Early studies determined that status and role characteristics of individuals receiving warning messages influence protective action initiation time. It has also been found that those who are younger,[33] have attained higher levels of education,[34] and are employed,[35] in addition to women as compared to men,[36] are more likely to interpret alert and warning information better and take appropriate protective actions.

RECENT RESEARCH

A number of recent studies have yielded additional insights building on the decades of earlier research on disaster response.

J. Nehnevajsa, 1985, *Western Pennsylvania: Some Issues in Warning the Population Under Emergency Conditions*, Pittsburgh, PA: University Center for Social and Urban Research, University of Pittsburgh; G. Rogers and J. Sorensen, 1991, Diffusion of emergency warning: Comparing empirical and simulation results, *Risk Analysis* 11:117-134.

[33] S. Cutter and K. Barnes, 1982, Evacuation behavior and Three Mile Island, *Disasters* 6(2):116-124; H. Friedsam, 1962, Older persons in disaster, pp. 151-184 in *Man and Society in Disaster* (G. Baker and D. Chapman, eds.), New York, NY: Basic Books; R. Mack and G. Baker, 1961, *The Occasion Instant: The Structure of Social Responses to Repeated Air Raid Warnings*, Disaster Study No. 15, Washington, DC: National Research Council, National Academy of Sciences; R. Perry, M. Lindell, and M. Greene, 1981, *Evacuation Planning in Emergency Management*, Lexington, MA: Lexington Books; B. Phillips and B. Morrow, 2007, Social science research needs: Focus on vulnerable populations, forecasting, and warnings, *Natural Hazards Review* 8(3):61-68.

[34] T. Drabek, 1986, *Human System Responses to Disaster: An Inventory of Sociological Findings*, New York, NY: Springer Verlag; D. Mileti, and C. Fitzpatrick, 1993, *The Great Earthquake Experiment: Risk Communication and Public Action*, San Francisco, CA: Westview Press; R. Turner, J. Nigg, D. Paz, and B. Young, 1979, *Earthquake Threat: The Human Response in Southern California*, Los Angeles, CA: Institute for Social Science Research, University of California.

[35] C. Flynn, 1979, *Three Mile Island Telephone Survey: Preliminary Report on Procedures and Findings*, Tempe, AZ: Mountain West Research; R. Lachman, M. Tatsuoka, and W. Bonk, 1961, Human behavior during the tsunami of May, 1960, *Science* 133:1405-1409; R. Perry, 1987, Disaster preparedness and response among minority citizens, *Sociology of Disasters* 135-151; R. Stallings, 1984, Evacuation behavior at Three Mile Island, *International Journal of Mass Emergencies and Disasters* 2:11-26; Y. Yamamoto and E. Quarantelli, 1982, *Inventory of the Japanese Disaster Literature in the Social and Behavioral Sciences*, Columbus, OH: Disaster Research Center, Ohio State University.

[36] T. Drabek, 1986, *Human System Responses to Disaster: An Inventory of Sociological Findings*, New York, NY: Springer Verlag; D. Mileti, and C. Fitzpatrick, 1993, *The Great Earthquake Experiment: Risk Communication and Public Action*, San Francisco, CA: Westview Press; and R. Turner, J. Nigg, D. Paz, and B. Young, 1979, *Earthquake Threat: The Human Response in Southern California*, Los Angeles, CA: Institute for Social Science Research, University of California.

Communicating Time Until Impact

WEA messages contain several elements—hazard, location, source, guidance, and time until impact. The Study of Terrorism and Responses to Terrorism (START) research team[37] found that both *guidance* (what to do and how to do it) and *time until impact* (how much time people have to take the recommended action) play major roles relative to other message elements in the outcomes of public understanding and belief of the protective action recommendation and the ability to decide how to respond. Importantly, the START research team found that WEA messages would be more effective if they were to state how much time remains until impact rather than use time to indicate when the message expires, as is the current practice.

WEA messages are designed to provide alerts about imminent threats; the START research team conceptualized imminent as occurring within one hour. Other research has extensively examined the optimal timing of warnings for other alerting systems. For example, research on tornados finds that the optimal lead for issuing a tornado warning is from 15 minutes to just over 30 minutes.[38] If too much lead time is provided, people are less likely to follow the protective guidance in a timely manner. Therefore, understanding how to best express lead time in WEA messages and the ideal lead times by hazard type are important areas for future research.

Including Protective Guidance in Web Links

Prior research called for including URLs in WEA messages to provide access to more complete information on the hazard and recommended protective guidance.[39] Christopher McIntosh from Esri, a geographical information system firm, observed in his testimony to the committee that "without context, alerts are just noise."

At this point, it is unclear what information is best included in a WEA message and what information is best included in Web pages the mes-

---

[37] M. Wood, H. Bean, B. Liu, and M. Boyd, 2015, *Comprehensive Testing of Imminent Threat Public Messages for Mobile Devices: Final Report*, College Park, MD: National Consortium for the Study of Terrorism and Responses to Terrorism.

[38] S. Hoekstra, R. Butterworth, K. Klockow, D.J. Drotzge, and S. Erickson, 2011, A social perspective of warn on forecast: Ideal tornado warning lead time and the general public's perceptions of weather risks, *Weather, Climate & Society* 3(1):128-140; and K.M. Simmons and D. Sutter, 2009, False alarms, tornado warnings, and tornado casualties, *Weather, Climate & Society* 1(1):38–53.

[39] M. Wood, H. Bean, B. Liu, and M. Boyd, 2015, *Comprehensive Testing of Imminent Threat Public Messages for Mobile Devices: Final Report*, College Park, MD: National Consortium for the Study of Terrorism and Responses to Terrorism.

sage links to. Prior research on WEA did not examine 360-character messages because the research was conducted before the Federal Communications Commission rulemaking that extended WEA messages from 90 to 360 characters.[40] Furthermore, concerns remain as network congestion could be caused by people accessing an included link within seconds of receiving a WEA that includes a URL.[41] Interestingly, some research found that message recipients are unlikely to open linked content and that instead they read only a few words of WEA-like messages owing to stress responses.[42] Therefore, research is needed to understand how to most effectively craft 360-character messages as well as under what circumstances and what message content should be included in linked media. Also unknown is how to craft WEAs so that they galvanize people to read the entire message, including potentially life-saving linked content. Research is needed on how to best convey protective action guidance in 360-character messages vs. linked media. A consistent finding across research on WEA messages is that the public needs education on what the WEA service is as well as what protective actions to take during a variety of hazards.

Geotargeting

More precise geotargeting that leverages the information that smartphones have about one's location could be used to deliver more accurate and relevant alerts. Research has shed some light on the possible impacts of geotargeting on PAI time.

Some research has looked at the impact of reducing the size of the zone receiving an alert.[43] It found that individuals receiving these more targeted messages would nevertheless forward these messages to individuals outside the zone. In some cases, this may decrease PAI time because people receive an alert from additional sources. In other cases, people may end up receiving forwarded messages intended for different zones that may call for the wrong protective action. For example, someone

---

[40] Federal Communications Commission, "FCC Strengthens Wireless Emergency Alerts as a Public Safety Tool," release date September 29, 2016, https://apps.fcc.gov/edocs_public/attachmatch/DOC-341504A1.pdf.

[41] Federal Communications Commission, "Improving Wireless Emergency Alerts and community-initiated alerting," release date November 19, 2015, https://apps.fcc.gov/edocs_public/attachmatch/FCC-15-154A1.pdf.

[42] D. Glik, University of California, Los Angeles, "WEA Messages: Impact on Physiological, Emotional, Cognitive and Behavioral Responses," presentation to committee on September 1, 2016.

[43] A. Parker, RAND Corporation, "Exploring the Effect of the Diffusion of Geo-Targeted Emergency Alerts," presentation to the committee on September 1, 2016.

intended to get a message to shelter in place may receive a forwarded message to evacuate.

Other research has looked at the location-proxy fallacy—that only those in the alert zone area would be interested in received messages for those zones.[44] In fact, alerts may be of value to not only people in the alert zone at a specific time but also those who are contemplating entering or frequently enter that zone.[45] Furthermore, individuals outside the geo-targeted area may share information with those at risk, adding credibility to the alert.

Constructing geotargeting algorithms that use cell site propagation footprints has been found to be the best method for sending out these messages as it can result in much smaller areas regardless of the physical location of cell towers and improve granularity, allows for monthly tests of the system without impacting the general public, enhances the reachability to people in harm's way, and requires no change to the current WEA network.[46] Another method of distributing these messages is a mechanism called Arbitrary-Size Location-Aware Targeting (ASLAT). Using this technique does not consume excessive mobile device power or radio resources, but increases geotargeting accuracy by utilizing the location awareness of mobile devices. It also increases delivery time as the phone will know its location before processing a received alert and will be able to determine if a certain message needs to get to the person or not. Implementation of ASLAT would require some changes to existing WEA standards for specific functionality in the cellular networks and mobile device behaviors.

Additionally, the use of geotargeting could address the issue of over-alerting. In one study, two different methods were used to distribute WEA messages in order to calculate geotargeting performance (GTP) estimates under two imminent threat scenarios: tornado warnings and earthquake early warning. In the first method, only cell towers within the warning area were directed to broadcast WEA messages, while in the second method WEA messages were broadcasted by cell towers within the warning area as well as towers adjacent to the warning area. It was found that over-alerting rates were lower when using the first method, but that if using alert failure rate (AFR) as the primary metric of GTP, the second method provided superior GTP in urban and mixed areas.[47] The

---

[44] B. Iannucci, Carnegie Melon University, "Opportunities, Options, and Enhancements for the Wireless Emergency Alerting Service," presentation to the committee on September 1, 2016.

[45] A. Parker, RAND Corporation, "Exploring the Effect of the Diffusion of Geo-Targeted Emergency Alerts," presentation to the committee on September 1, 2016.

[46] D. Ung, TeleCommunication Systems, Inc., "Geo-Targeting Method Using Cell Radio Frequency (RF) Propagation," presentation to the committee on September 1, 2016.

[47] D. Gonzales, 2016, *Geo-Targeting Performance of Wireless Emergency Alerts in Imminent Threat Scenarios – Volume 1: Tornado Warnings*, Washington, DC: Department of Homeland Security.

study noted that for tornados, the warning polygon is fixed for hours, although the path of the tornado may change. Initiatives like Threats In Motion (TIM) may be of value, since the tornado warning geotargeting would be improved by updating the portion of the warning polygon more rapidly. However, it is acknowledged that trying to integrate this system into WEA may present many challenges, including the transmission of more WEA messages and testing would be needed to ensure that WEA could handle TIM-based tornado warnings.[48]

Message Delivery Method

How an individual receives an alert and warning message can influence an individual's perception on the risk and threat and therefore affects their PAI time. Earlier research on alert and warnings messages could not, of course, have factored in the opportunities and challenges presented by the recent dramatic changes in how people receive information. There are now a variety of new ways that an alert and warning message can reach an individual, including WEA messages, social media, phone applications, and online messaging among friends and family. Together with traditional media, such as television and radio, these constitute a complex, evolving ecosystem with many interacting systems.

Social Media

There are unexplored opportunities to utilize social media as complementary channels for emergency alerts[49]—including uses related to both incoming and outgoing information. This aligns with a "ubiquitous alerting" strategy[50] that includes all available channels and devices.

In the United States, 69 percent of adults use some type of social media.[51] Social media are widely used during disaster events by emergency responders, people in the affected community, and global onlookers[52]

---

[48] Ibid.
[49] Communications, Security, Reliability and Interoperability Council V (CSRIC V), Working Group 2, Emergency Alerting Platforms, 2016, *Social Media & Complementary Alerting Methods – Recommended Strategies & Best Practices: Final Report & Recommendations*, Washington, DC: Federal Communications Commission.
[50] R. Wimberly, "New Age of Alerting Coming: Ubiquitous Alerts," release date February 9, 2016, http://www.emergencymgmt.com/emergency-blogs/alerts/new-age-of-alerting-coming--ubiquitous-alerts.html.
[51] Pew Research Center, "Social Media Fact Sheet," release date January 12, 2017, http://www.pewinternet.org/fact-sheet/social-media/.
[52] A.L. Hughes and L. Palen, 2009, Twitter adoption and use in mass convergence and emergency events, *International Journal of Emergency Management* 6(3-4):248-260.

who converge there to seek and share information.[53] While social media platforms can be used to gather and disseminate information, it can also be leveraged to help organize response efforts[54] and share messages of support.[55] Researchers have noted the potential for these platforms to contribute (as an incoming information source) to enhanced situational awareness of emergency responders and affected community members[56]—by aggregating information from distributed users who may have access to different perspectives of the disaster events. There is also opportunity for these platforms to be used as real-time communication tools for official responders to distribute messages to their various publics—and indeed many alert originators (including several regional offices within the National Weather Service) are already utilizing social media as part of their alerting activities.[57]

However, there are also many challenges related to the use of social media in the crisis context. For those monitoring social media during disaster events for safety-critical and/or actionable information, it can be difficult to identify the signal from within the noise, due to the volume of information shared. Significant for conversations about social media as an emergency alerting platform, emergency responders face several concerns regarding use of social media:

---

[53] L. Palen and S.B. Liu, 2007, Citizen communications in crisis: Anticipating a future of ICT-supported public participation, pp. 727-736 in *Proceedings of the SIGCHI Conference on Human Factors in Computing Systems*, https://dl.acm.org; J.N. Sutton, L. Palen, and I. Shklovski, 2008, Backchannels on the front lines: Emergent uses of social media in the 2007 Southern California wildfires, pp. 624-632 in *Proceedings of the 5th International ISCRAM Conference*, http://www.iscramlive.org/portal/node/2236.

[54] K. Starbird and L. Palen, 2011, Voluntweeters: Self-organizing by digital volunteers in times of crisis, pp. 1071-1080 in *Proceedings of the SIGCHI Conference on Human Factors in Computing Systems*, https://dl.acm.org; H. Gao, G. Barbier, and R. Goolsby, 2011, Harnessing the crowdsourcing power of social media for disaster relief, *IEEE Intelligent Systems* 26(3):10-14; J.I. White and L. Palen, 2015, Expertise in the wired wild West, pp. 662-675 in *Proceedings of the 18th ACM Conference on Computer Supported Cooperative Work & Social Computing*, https://dl.acm.org.

[55] K. Starbird and L. Palen, 2012, (How) will the revolution be retweeted? pp. 7-16 in *Proceedings of the ACM 2012 conference on Computer Supported Cooperative Work*, https://dl.acm.org.

[56] S. Vieweg, A.L. Hughes, K. Starbird, and L. Palen, 2010, Microblogging during two natural hazards events: What Twitter may contribute to situational awareness, pp. 1079-1088 in *Proceedings of the SIGCHI Conference on Human Factors in Computing Systems*, https://dl.acm.org.

[57] Communications, Security, Reliability and Interoperability Council V (CSRIC V), Working Group 2, Emergency Alerting Platforms, 2016, *Social Media & Complementary Alerting Methods – Recommended Strategies & Best Practices: Final Report & Recommendations*, Washington, DC: Federal Communications Commission.

- They are challenged to keep up with the volume and to find relevant and actionable information.
- Social media platforms have opened up two-way channels of communication with the public. Members of the public expect to be heard on social media—i.e., for emergency officials to respond to what members of the public share on social media.[58]
- Though several early adopters have had some success with utilizing these platforms, the "rules for engagement" and best practices for emergency responders using social media are still evolving.[59]
- Social media are not geographic-specific and therefore it is difficult for responders to distinguish between their local community members and the global audience.
- Emergency responders fear the spread of misinformation on these platforms, and many have reported a reluctance to adopt social media in part due to this concern.[60,61]

Another challenge, which suggests the need for more research, is the varied use of different social media platforms across different demographics, a problem that could create new "digital divides" in the accessibility of information if alerts are shared via social media.[62] This latter point suggests an "all channels" strategy, which positions social media in general as a complementary alerting source and not a primary one, and points toward the use of many different platforms (e.g., Twitter, Facebook, and Snapchat) at once.

Until recently, little was known about the relative effectiveness in prompting protective action or complementarity of alerts distributed through WEA or social media platforms, such as Twitter, which has been used by emergency managers since its debut in 2009.[63] In one study,

---

[58] S. Bernier, CEM, CBCP, MBCI SB Crisis Consulting, 2013, "Social Media and Disasters: Best Practices and Lessons Learned," presentation at the Disaster Preparedness Summit, August 21, American Red Cross, Chicago, IL.

[59] K. Starbird, D. Dailey, A.H. Walker, T.M. Leschine, R. Pavia, and A. Bostrom, 2015, Social media, public participation, and the 2010 BP Deepwater Horizon oil spill, *Human and Ecological Risk Assessment: An International Journal* 21(3):605-630.

[60] S.R. Hiltz, J. Kushma, and L. Plotnick, 2014, Use of social media by U.S. public sector emergency managers: Barriers and wish lists, pp. 602-611 in *Proceedings of the 11th International ISCRAM Conference*, http://www.iscramlive.org/portal/node/2236.

[61] A.L Hughes and L. Palen, 2012, The evolving role of the public information officer: An examination of social media in emergency management, *Journal of Homeland Security and Emergency Management* 9(1):1-20.

[62] K. Crawford and M. Finn, 2015, The limits of crisis data: Analytical and ethical challenges of using social and mobile data to understand disasters, *GeoJournal* 80(4):491-502.

[63] A.L. Hughes and L. Palen, 2009, Twitter adoption and use in mass convergence and emergency events, *International Journal of Emergency Management* 6(3-4):248-260.

which used mock WEA messages (90 characters) and mock Twitter-length messages (140 characters), the most common initial reaction was confusion and fear accompanied by a desire to acquire more information about the apparent threat.[64] The lack of clarity about the hazard, the protective action guidance and how long they had to complete it, the time of the incident, and the affected area led to confusion and frustration. Because of the limited information and lack of personalization in these messages, most participants in the study could not tell whether the hazard would impact them specifically.[65] Consistent with public warning response models,[66] participants would have sought out additional information before taking any protective action.

Like most technology, social media platforms are continuously evolving, as are the practices related to their use, both generally and in the context of emergency alerting and response. Thus, there is a need to continue research in this area to better understand changing practices in order to more fully realize the acknowledged potential of these platforms. Specifically, there are unexplored opportunities to utilizing social media as complementary channels for emergency alerts—including uses related to both incoming and outgoing information. This is especially important since the research record provides strong evidence that how warning message(s) are delivered to the public influences public warning message response (referred to as warning or message delivery factors). Message delivery factors can help people confirm the warning message and personalize risk, for example, which are both important intervening factors between getting a message and taking a recommended protective action. It has been found that messages frequently repeated work best to decrease PAI time, and messages delivered over multiple and different communication channels work best in comparison to those delivered over fewer or single channels. In the context of social media, familiar accounts (e.g., friends, family, and celebrities), have the ability to share emergency alert and warning messages from official accounts, which adds to the credibility of the message. However, this can also lend itself to adding credibility to incorrect or misleading information.

---

[64] H. Bean, B.F. Liu, S. Madden, J. Sutton, M.M. Wood, and D.S. Mileti, 2016, Disaster warnings in your pocket: How audiences interpret mobile alerts for an unfamiliar hazard, *Journal of Contingencies and Crisis Management* 24(3):136-147.

[65] Ibid.

[66] D. Mileti and J. Sorensen, 1990, Communication of Emergency Public Warnings: A Social Science Perspective and State-of-the-Art Assessment, Oak Ridge, TN: Oak Ridge National Laboratory, U.S. Department of Energy; R. Perry, 1979, Evacuation decision-making in natural disasters, *Mass Emergencies* 4:25-38.

Crowdsourcing Technologies

Crowdsourcing tools, which include social media platforms such as Twitter and navigation services such as Waze,[67] could leverage the capacity of the online crowd to help process and distribute emergency messages. While emergency messages disseminated through crowdsourcing technologies may not be more accurate than government emergency alerts themselves, they may be a trusted source because they are used on a day-to-day basis. Such has been the experience of Humanity Road, a virtual volunteer organization that assists with humanitarian response and disaster preparedness education. Humanity Road takes information from a variety of sources,[68] analyzes it, and standardizes it for dissemination on Twitter. It developed what it calls Twitter Commandments (Box 1.3) that reflect lessons learned on how to most effectively collect, analyze, standardize, and disseminate disaster information.[69]

The crowdsourced navigation service Waze collects data, including traffic conditions and road hazards from its users, to generate optimal routes. To combat misinformation, user reports have to be validated by other users. Apps like Waze can be used during events to collect and distribute information about hazards such as road closures, floods, and accidents. Moreover, this information can be supplemented by alert originators who can transmit updated information on hazards or suggest safe travel routes as information is gathered throughout an event. In Texas, the Center for Collaborative Adaptive Sensing of the Atmosphere (CASA) Dallas Fort Worth Living Lab has already begun to experiment with how to provide information about urban flooding hazards.[70] The CASA Alerts Application introduced personalization and allowed users to control thresholds and preferences, which determined which alerts reach the user. Users, in turn, can provide feedback by verifying data related to the event, providing information about their experience during the event,

---

[67] While Waze relies heavily on volunteer information, reported incidences are validated through other users. Additionally, during a disaster much of its information (e.g., shelters) is confirmed by official sources.

[68] Sources used by Humanity Road include government sources, although it is noted that there is an inherent distrust of the government among a large portion of the population.

[69] K. Starbird and L. Palen, 2013, "Working & Sustaining the Virtual 'Disaster Desk'," University of Washington, http://faculty.washington.edu/kstarbi/cscw2013_final-2.pdf.

[70] The CASA Dallas Forth Worth Living Lab brings together the NCTCOG, NWS, DFW Airport, Fort Worth, Grand Prairei, Midlothian, and other public safety officials, in addition to bringing in the user. They do experiments in the community rather than doing experiments in a lab setting (E.J. Bass, B. Hogan, D. Rude, C. League, P. Marsh, L. Lemon, B. Philips, D. Westbrook, J. Brotzge, and R. Riley, 2011, "A method for investigating real-time distributed weather forecaster-emergency manager interaction," pp. 2809-2815 in *2011 IEEE International Conference on Systems, Man, and Cybernetics*, doi: 10.1109/ICSMC.2011.6084098.).

> **BOX 1.3**
> **Humanity Road Twitter Commandments for Crisis Tweeter**
>
> 1. Do no harm. It is safer to share no news than to share inaccurate news. Rumors put lives at risk.
> 2. Do not panic. You do not need to know it all.
> 3. Take a deep breath. Do not be distracted by noise & confusion.
> 4. Verify source. If you do not know for sure the source is reliable – do not RT [retweet] the information. Use 2 reliable independent sources for major news.
> 5. Verify facts. Get facts, locations, address, specific need, # of people impacted, population at risk, dig deep into details, the more the better.
> 6. Listen to the official & experts. Use caution & reason & follow those who appear to have a "handle" on how to respond in these situations.
> 7. Use tweak the tweet. We recommend formatting your tweet using Tweak the Tweet.
> 8. Not sure where to start? Pick one topic and stick with it. Become an expert on fielding, research facts and providing help on specific topic.
> 9. Repeat the first 3 Twitter Commandments as needed.
>
> SOURCE: Reprinted from K. Starbird and L. Palen, 2013, "Working & Sustaining the Virtual 'Disaster Desk'," University of Washington, http://faculty.washington.edu/kstarbi/cscw2013_final-2.pdf.

and providing information on what action, if any, they took during the event. By getting feedback from users, more information can be gathered relating to what areas are being impacted and how, and gives researchers the ability to gather information on how many individuals took the recommended protective action(s).

Message Receiver Characteristics

Additional factors that could influence an individual's PAI time are an individual's characteristics. These characteristics include an individual's age, gender, ethnicity, socioeconomic status, race, and disability status.

Age

A recent study looking at the effects of weather on activities by the elderly using the University of Michigan-Thomson Reuters Surveys of Consumers helps shed some light on the vulnerability of the elderly in adverse conditions. This study found that icy conditions were the hazard type that most caused participation restriction (where people took the protective action of not going outside or engaging in certain activities) in day-to-day life,[71] as compared to cold temperatures, snow, rain, fog, and wind.[72] Adults 65 or older reported greater difficulty leaving the home in icy conditions, especially in cases where these adults had mobility impairment or self-imposed driving restrictions.[73] Other work has found that, owing to higher rates of chronic illnesses, older persons are more susceptible to the adverse effects of psychological and physical stress, including disasters.[74] Although there are resources and guidelines for older adults, those with mobility impairment continue to have difficulties both during and after a disaster, and the older an individual is, the less they are prepared during and after a disaster. Another study, which sampled adults aged 50 and above from the Health and Retirement Study, found that two-thirds of the study population lacked an emergency plan, had never participated in a preparedness program, and were not aware of the resources available to them.[75] In addition, 15 percent of respondents reported having medical equipment that needed electricity, making them susceptible to power outages.[76] An additional problem with older adults is that social isolation may prevent them from receiving warning messages or asking for help when needed, which may render them invisible to rescue teams during times of disasters.[77] The problem of social isolation is compounded by differences in mobile device ownership, where a small subset of elderly individuals do not own a cell phone or they own a device that is not a smartphone (which impacts the types of messages they can receive). In the United States, 97 percent of older adults between the ages of 50 and 64 and 80 percent of older adults aged 65 and above were found to own a mobile device. In the group between the ages of 50 and 64, 74 percent owned smartphones while that percent dropped to 42 percent for those 65 or older.[78] In an effort to mitigate the issue of social isolation, the Department of Health and Human Services has created the emPOWER map tool, which maps multiple populations with various

---

[71] This included participation in grocery shopping, playing sports, and driving.

[72] P.J. Clarke, T. Yan, F. Keusch, N.A. Gallagher, 2015, The impact of weather on mobility and participation in older U.S. adults, *American Journal of Public Health* 105(7):1489-1494.

[73] Ibid.

[74] T.M. Al-rousan, L.M. Rubenstein, and R.B. Wallace, 2014, Preparedness for natural disasters among older US adults: A nationwide survey, *American Journal of Public Health* 104(3):506–511.

[75] Ibid.

[76] Ibid.

[77] D.P. Eisenman, K.M. Cordasco, S. Asch, J.F. Golden, and D. Glik, 2007, Disaster planning and risk communication with vulnerable communities: Lessons from Hurricane Katrina, *American Journal of Public Health* 97(Suppl 1):S109-S115.

[78] Pew Research Center, "Mobile Fact Sheet," release date January 12, 2017, http://www.pewinternet.org/fact-sheet/mobile/.

kinds of vulnerabilities that could potentially be used to identify vulnerable, at-risk populations during disasters.[79]

Gender, Ethnicity, and Socioeconomic Status

Those perceived to be the most vulnerable during natural disasters are older adults, women, and minorities. While looking at perceptions and circumstances by gender and ethnicity, a study conducted in Rhode Island found that women were more likely to have children or rent a home, while being less likely to earn an income of $100,000 or greater as compared to men. Minority populations, in comparison to white men and women, often have nowhere to stay if a hurricane hits, and are less likely to own a car or have an income greater than $100,000. It was found that it is not always the case that women and minority populations act differently than men or white men and women, respectively, but what influences perceptions of hazard (as subject to race and gender effects) are a person's financial resources, home/car ownership, and relationships (marriage or children).[80] Additionally, resilience is hindered for those with limited social connectedness.[81]

This emphasis on relationships, specifically within minority communities, was seen following Hurricane Katrina. In Houston's major evacuation center, 58 qualitative interviews were conducted with individuals—most of whom were from New Orleans, low income, and African American. The study concluded that overcoming shelter and transportation obstacles would not have been sufficient to significantly improve disaster response. More important were a person's ties to family, friends, and the community, suggesting that disaster preparedness strategies emphasize better community-based communications.[82]

Language and Culture

The U.S. population is becoming increasingly racially and ethnically diverse; it is predicted that by 2055, the United States will not have a

---

[79] Department of Health and Human Services, "HHS emPOWER Map 2.0," last updated December 30, 2016, https://empowermap.hhs.gov/.

[80] D.M. West, and M. Orr, 2007, Race, gender, and communications in natural disasters, *Policy Studies Journal* 35(4):569–586.

[81] D.P. Aldrich, 2014, *Building Resilience: Social Capital in Post-Disaster Recovery*, Chicago, IL: University of Chicago Press, August.

[82] D.P. Eisenman, K.M. Cordasco, S. Asch, J.F. Golden, and D. Glik, 2007, Disaster planning and risk communication with vulnerable communities: Lessons from Hurricane Katrina, *American Journal of Public Health* 97(Suppl 1):S109-S115.

single racial or ethnic majority.[83] The resulting linguistic and cultural diversity can present a challenge for emergency managers who may not have the resources to communicate effectively with these populations. For example, in New York City, there are 208 spoken languages, but emergency alert and warning messages are available only in 18 languages.[84] This leaves a large portion of the population susceptible to not receiving or to misunderstanding WEA messages. A study conducted in three Gulf Coast counties in South Mississippi with highly diverse communities[85] found that 47.3 percent of respondents indicated they wished to receive alert and warning messages in a language other than English. Although there are a variety of technologies (further discussed in Chapter 3) that are available to disseminate alert and warning messages, including WEA, TV, and radio, it was found that authorities had to depend on posters and pictures to disseminate information to Vietnamese and Hispanic communities.[86] The respondents to the study indicated that they trust and are more likely to act on information received from their family and friends. The second trusted source of information was information received via TV, radio, and WEA. Respondents in the study were also more likely to trust sirens than social media.[87]

Many regions in the United States have large populations for whom English is not the primary language, and the number of different languages spoken in some areas can be very large. However, even as language diversity is increasing, the use of smartphones and other technologies (e.g., Internet, broadband, tablet, and social media), which provide valuable translation capabilities, is also growing (Table 1.1). Smartphone adoption has more than doubled since 2011, when the Pew Research Center started surveying the topic.[88] With smartphones, individuals (assuming

---

[83] D. Cohn and A. Caumont, "10 Demographic Trends That Are Shaping the U.S. and the World," Pew Research Center, release date March 31, 2016, http://www.pewresearch.org/fact-tank/2016/03/31/10-demographic-trends-that-are-shaping-the-u-s-and-the-world/.

[84] B.J. Krakauer, New York City Emergency Management, "Vision for the Future of Public Alerts and Warning: New York City's Perspective," presentation to the committee on August 9, 2016.

[85] Hancock, Harrison, and Jackson counties in Mississippi. The Mississippi Gulf Coast is susceptible to natural disasters like tropical cyclones and coastal flooding, in addition to being a landfall location for Hurricane Katrina. These highly populated counties have culturally diverse communities, including Anglo-Americans, African Americans, and Vietnamese and Hispanic immigrants.

[86] B. Kar, University of Southern Mississippi, "An Integrated Approach to Geo-Target At-Risk Communities and Deploy Effective Crisis Communication Approaches," presentation to the committee on September 1, 2016.

[87] Ibid.

[88] A. Smith, "Record Shares of Americans Now Own Smartphones, Have Home Broadband," Pew Research Center, release date January 12, 2017, http://www.pewresearch.org/fact-tank/2017/01/12/evolution-of-technology/.

TABLE 1.1 Cell Phone Ownership

	Owns Any Cell Phone	Owns Smartphone	Owns Cell Phone That Is Not a Smartphone
Income			
Less than $30,000	92%	64%	29%
$30,000-$49,999	95%	74%	21%
$50,000-$74,999	96%	83%	13%
$75,000+	99%	93%	6%
Race/Ethnicity			
White	94%	77%	17%
Black	94%	72%	23%
Hispanic	98%	75%	23%

SOURCE: Pew Research Center, "Mobile Fact Sheet," release date January 12, 2017, http://www.pewinternet.org/fact-sheet/mobile/.

they know how to use the phone and have a data plan or WiFi access) have many tools at their disposal. For example, Google Translate could be used to translate an alert into a person's native language, and Facebook has an integrated translation capability. However, despite rapid improvements in recent years, such tools are imperfect and may in some cases cause further confusion. A particular challenge is that such tools do not account for differences in usage among different dialects.

Culture can also affect how people respond to various states of a disaster. For example, the October 17, 1989, Loma Prieta earthquake caused extensive damage in Watsonville, Santa Cruz, and Los Gatos, California. Families with experience with earthquakes in Mexico preferred to camp outdoors rather than stay in possibly damaged buildings, yet it took time for authorities to agree to open public parks as official shelters. Those camping in the parks were also motivated by a desire to stay close to their homes in order to protect personal possessions.[89] The challenge of sheltering all the disaster victims was further compounded by the lack of preplanning for the situation, which stemmed from a lack of community participation during the preplanning despite prior research calling for such engagement.[90] Another cultural issue arose regarding shelters when victims who were refugees from Central America found the tents and fences set up by the

---

[89] B.D. Phillips, 1993, Cultural diversity in disasters: Sheltering, housing, and long term recovery, *International Journal of Mass Emergency Disasters* 11(1):99-110.
[90] E.L. Quarentelli, K. Green, E. Ireland, S. McCabe, and D.M. Neal, 1983, *Emergent Citizen Groups in Disaster Preparedness and Recovery Activities*, Columbus, OH: Ohio State University.

American Red Cross and National Guard too reminiscent of concentration camps and government-backed death squads. These differences in perceptions highlight the importance of involving the community in the creation, planning, and execution of emergency practices.

Disability

In a study conducted to understand how disability factors into the response to WEA messages, it was shown that individuals familiar with WEA were more likely to take immediate action,[91] but individuals with a disability are only half as likely to have heard of WEA. Like other populations, the overwhelming majority of those who participated in the study (98 percent) reported owning a mobile phone. As a result, these populations stand to benefit from using devices that provide appropriate attention-getting, sound, and display affordances—capabilities that can also help nondisabled populations.[92]

Physiological and Mental Models

When it comes to forwarding or sharing messages, the type of hazard and associated biases affect what individuals share. One study found that subjects (shown both WEA and Twitter messages) were more predisposed to share dismissive messages and tweets about floods as compared to blizzards, gas leaks, hurricanes, and tornados.[93] When deciding to share a message, there seemed to be a more deliberate thought process, as the subjects in the study typically had a higher level of brain activity in the frontal lobe compared to when they chose not to share a message. The study showed that, overall, subjects shared disaster messages a majority of the time and were highly responsive to all types of disaster messages (WEA messages and tweets).[94] However, the study also suggested that there are perceived differences in the threat or urgency posed by various disasters on a physiological level, which explains why the subjects reacted to floods differently than the other hazards.[95] With certain hazards (such as floods) there seems to be an optimism bias, in which a person believes that the probability

---

[91] H. Mitchell, "Optimizing Wireless Emergency Alerts for Sensory Disabilities," presentation to the committee on September 1, 2016.
[92] Ibid.
[93] C.D. Corley, N.O. Hodas, R. Butner, J.J. Harrison, and C. Berka, 2016, "Modeling Cognitive Response to Wireless Emergency Alerts to Inform Emergency Response Interventions," Pacific Northwest National Laboratory, https://www.dhs.gov/sites/default/files/publications/WEA%20-%20Modeling%20Cognitive%20Response.pdf.
[94] Ibid.
[95] Ibid.

of the event actually happening is low, that influences their decision of whether or not to engage in the recommended, protective actions.

Alert Originators

Part of increasing public participation in protective actions is educating the public about alert and warning systems. Although the Federal Emergency Management Agency (FEMA) partnered with the Ad Council to create an education campaign about WEA, there seems to be little knowledge among the public about what WEA is, what its purpose is, or how and why alerts are issued.[96] Indeed, participants from a variety of studies have questioned the validity of the emergency alerts they have received. In addition, some participants do not know the meaning of acronyms used in messages[97] and in one study, participants asserted that information from local sources would be more believable than WEA alerts.[98] Education of the public is just one of the avenues that may be required to enhance the effectiveness of alert and warning messages.

Alert originators also need to be knowledgeable about the capabilities and shortcomings of the WEA systems.[99] This can include understanding WEA local area coverage, as cellular networks can vary from one region to another, especially in rural areas.[100] In order to provide alert originators with information concerning the implementation and utilization of WEA, the Software Engineering Institute published papers on integration,[101] best practices,[102] and security.[103] Alert originators also need to know how to write effective messages under the time crunch of emergencies.

---

[96] H. Bean, B.F. Liu, S. Madden, J. Sutton, M.M. Wood, and D.S. Mileti, 2016, Disaster warnings in your pocket: How audiences interpret mobile alerts for an unfamiliar hazard, *Journal of Contingencies and Crisis Management* 24(3):136-147.

[97] Ibid.

[98] Ibid.

[99] C. Woody, Software Engineering Institute of Carnegie Mellon University, "SEI Wireless Emergency Alerts (WEA) Research 2013 through 2016," presentation to the committee on September 1, 2016.

[100] D. Gonzales, 2016, *Geo-Targeting Performance of Wireless Emergency Alerts in Imminent Threat Scenarios–Volume 1: Tornado Warnings,* Washington, DC: Department of Homeland Security.

[101] Carnegie Mellon University, 2014, *Commercial Mobile Alert Service (CMAS) Alerting Pipeline Taxonomy,* CMU/SEI-2013-TR-019, Pittsburgh, PA.

[102] Carnegie Mellon University, 2013, *Best Practices in Wireless Emergency Alerts,* CMU/SEI-2013-SR-015, Pittsburgh, PA; Carnegie Mellon University, 2013, *Wireless Emergency Alerts New York City Demonstration,* CMU/SE I-2013-SR-tbd, Pittsburgh, PA; Carnegie Mellon University, 2014, *Maximizing Trust in the Wireless Emergency Alerts (WEA) Service,* CMU/SE I-2013-SR-027, Pittsburgh, PA; Carnegie Mellon University, 2014, *Wireless Emergency Alerts: Trust Model Technical Report,* CMU/SE I-2013-SR-021, Pittsburgh, PA.

[103] Carnegie Mellon University, 2014, *Wireless Emergency Alerts (WEA) Cybersecurity Risk Management Strategy for Alert Originators,* CMU/SE I-2013-SR-018, Pittsburgh, PA; Carnegie

If new capabilities, such as more precise geotargeting are made available, alert originators will need to understand those capabilities and how to best use them. Alert originators can also learn from the experiences of nongovernmental organizations like Humanity Road and operators of commercial services like Waze that may provide valuable insights on how to use two-way communications to get real-time information about what is happening in their area. By understanding how various platforms work, including social media as previously discussed in this chapter, alert originators may be able to leverage them more effectively in times of need in order to disseminate alerts and warnings to the public.

Mellon University, 2013, *Best Practices in Wireless Emergency Alerts*, CMU/SEI-2013-SR-015, Pittsburgh, PA; Mapping WEA Security Requirements and Guidance to Cybersecurity Risk Mitigation Recommendations (delivered separately to DHS - some requirements are restricted); INCOSE Insight Essay, 2013, Evaluation of security risk for WEA alert originators using mission threads, *Insight* 16(2).

2

Building an Integrated Alert and Warning Ecosystem

Alert and warning systems have evolved over time to reflect new types of hazards or technologies, moving from various television and radio broadcast technologies to now include cell broadcast. (See Figure 2.1 for the evolution of emergency broadcasting and Appendix A for a longer history of alert and warning systems.). However, this evolution has occurred very slowly and has often stemmed from a major hazards event. Furthermore, alert and warning systems have not kept up with new technologies. For example, the 2006 Warning, Alert, and Response Network (WARN) Act prompted the first significant changes to national alerting systems since the mid-1990s.

In combination with Executive Order 13407,[1] the WARN Act created the Integrated Public Alert and Warning System (IPAWS) and Wireless Emergency Alerts (WEA) (then known as the Commercial Mobile Alert System). IPAWS unified the Emergency Alert System (EAS), the national warning system (NAWAS), the newly created WEA, and National Oceanic and Atmospheric Administration (NOAA) Weather Radio All Hazards into a one modern network. Additionally, IPAWS allows for alerts to be originated by various government organizations and officials at the federal, state, local, and tribal level, and allows a single message to be transmitted to the various alert platforms. An XML-based data format, the Common Alerting Protocol (CAP), standardizes alert data across threats, jurisdictions, and warning systems. The CAP data structure was defined

---

[1] Executive Order. No. 13407, 2006, pp. 1226–1228.

Broadcasting

1951 - 1963 CONELRAD	1963 - 1997 EBS	1997 - 2006 EAS	2006 IPAWS
Originally called the "Key Station System," the **CON**trol of **EL**ectromagnetic **RAD**iation (CONELRAD) was established in August 1951.	EBS was initiated to address the nation through audible alerts. It did not allow for targeted messaging.	EAS jointly coordinated by the FCC, FEMA and NWS.	IPAWS modernizes and integrates the nation's alert and warning infrastructure.
Participating stations tuned to 640 & 1240 kHz AM and initiated a special sequence and procedure designed to warn citizens.	System upgraded in 1976 to provide for better and more accurate handling of alert receptions.	Designed for President to speak to American people within 10 minutes.	Integrates new and existing public alert and warning systems and technologies
	Originally designed to provide the President with an expeditious method of communicating with the American Public, it was expanded for use during peacetime at state and local levels.	EAS messages composed of 4 parts: • Digitally encoded header • Attention Signal • Audio Announcement • Digitally encoded end-of-message marker	Provides authorities a broader range of message options and multiple communications pathways
			Increases capability to alert and warn communities of all hazards impacting public safety.

FIGURE 2.1 The Evolution of Emergency Broadcasting. SOURCE: Federal Emergency Management Agency.

so as to be backward compatible with existing alert formats while providing new capabilities, such as those needed for WEA, flexible geotargeting that can more narrowly target areas using GIS data,[2] multilingual and multiaudience messaging; phased and delayed effective times and expirations, enhanced message update and cancellation features, template support, digital encryption and signature capabilities, and facilities for digital images, audio, and video.

Mobile devices have become an integral part of people's lives, with 97 percent of American adults owning one and 90 percent of those owners "frequently" carrying their phones with them and "never" or "rarely" power off the devices completely.[3] Furthermore, almost 43 percent of adults live in homes without a landline. Further limiting the ways in which homes can be reached, one in five households no longer have cable television subscriptions,[4] potentially limiting the reach of live, local news. Not only can cell phones reach a large swathe of the population, for some it may be the only or best way to reach them during emergencies given declines in listening to or viewing live broadcasts, a drop in cable subscription rates, and a dramatic falloff in households with landline telephones. The National Weather Service, in a talk in early 2015,[5] provided the following several examples where WEA was credited by the media and members of the public for saving lives:

- Rose Hill, MS, tornado on July 24, 2014,
- Cape Charles, VA, severe weather on July 24, 2013,
- Illinois tornadoes on November 17, 2013,
- East Windsor, CT, tornado on July 1, 2013, and
- Elmira, NY, tornado on July 26, 2012.

WEA, which added alerts delivered to phones to IPAWS, added important capabilities to the national alerting system but does not take full advantage of the ability of mobile devices to process and make decisions about which messages to present based on user needs or contextual information the device has about the user and the environment. Nor does

---

[2] Older alert systems relied on Federal Information Processing Standard codes that were assigned by county and a few larger cities, making the finest grain of geotargeting alerts countywide.

[3] Pew Research Center American Trends Panel Survey, May 30-June 30, 2014; L. Rainie and K. Zickuhr, "Americans' Views on Mobile Etiquette," release date August 26, 2015, http://www.pewinternet.org/2015/08/26/americans-views-on-mobile-etiquette/.

[4] A. Pressman, 2016, More than one in five households has dumped the cable goliath, *Fortune*, http://fortune.com/2016/04/05/household-cable-cord-cutters/.

[5] National Oceanic and Atmospheric Administration, "Wireless Emergency Alerts," last update February 5, 2015, http://www.crh.noaa.gov/images/lbf/wxsafety/WEA/WEA_Update%2002052015.pdf.

it leverage Internet-based technologies such as social media platforms that could be used to deliver alerts. Furthermore, both private and public organizations have begun to take advantage of the large amounts of data they possess about users to detect events and provide alerts and warnings and other hazard-related information to their users.

NEED FOR AN INTEGRATED ALERT AND WARNING ECOSYSTEM

Currently, emergency alerting takes place across an information ecosystem that includes emergency responders and their alerting platforms as well as diverse channels of message delivery, distributed sensing devices, and feedback mechanisms. Emergency alerts are distributed directly to users over landline phones and, more recently, over mobile phones (through WEA). They are also broadcast through traditional channels, such as radio and television, and increasingly through social media. Emergency alerts and additional contextual information about events can be accessed online through the websites of response agencies, mainstream media outlets, and other websites. Individuals also receive alerts via various mobile applications and other digital tools. In the near future, emergency alerts could also be delivered through other Internet-connected devices, such as Amazon Alexa or Google Home.

The information ecosystem for alerting the public encompasses more than IPAWS. For example, a person who receives a WEA message may post information from that alert on social media, or a community radio station may broadcast information it found in a Facebook post or another online source. The information ecosystem also includes "incoming" information. Information gathered from social media and distributed sensing devices can be utilized to inform situational awareness and generate emergency alerts—and (potentially) feedback mechanisms about what information is reaching whom, when, and how individuals are responding. Private organizations are also developing platforms to use during a crisis. Examples include Facebook Safety Check as a feedback channel and Google Alerts. The public also uses various commercial smartphone applications to provide weather alerting and follows local weather forecasters' social media feeds.

In addition to the technical changes, our understanding of how the public responds to systems has advanced. For example, while we have known for some time what information is needed to elicit public action, we also now know that the 90-character message length afforded by the current WEA system is not sufficient to yield a quick public response.[6] This ecosystem is continuing to evolve as new technologies are introduced and new practices and protocols emerge around information sharing during emergency events

The committee acknowledges the work done to develop and deploy current WEA capabilities and is encouraged by recent Federal Communications Commission (FCC) changes to WEA rules that expand the message length to 360 characters—changes made, at least in part, due to work funded by the Department of Homeland Security (DHS).[7] In view of the availability of new tools, and the emergence of Internet of Things (IoT) technologies, any methodology that relies on broadcasts to a singular device is no longer sufficient to serve as the primary alert and warning system for an increasingly connected population using diversified communication mediums and preferences.

> FINDING: Alert and warning systems exist within a larger communication and technical ecosystem, and government-designed and maintained systems must fit within this larger ecosystem.

> FINDING: A more cohesive and all-encompassing alert and warning system is needed that will integrate public and private communications mechanisms and sources of information, and continue to provide the necessary information for the purpose of preserving the health and safety of people, while being technologically agnostic—such that new technologies for alerts and warnings can be adopted quickly.

> FINDING: The nation's alerting capabilities, such as WEA and IPAWS, will need to evolve and progress as the capabilities of smartphones and other mobile broadband devices improve and newer technologies become available. This evolution will need to be informed by both technical research and social and behavioral science research.

PROPERTIES OF AN INTEGRATED ALERT AND WARNING SYSTEM

The purpose of alerts and warnings is to provide the necessary information to warn the public and effect the necessary actions that will lead

---

[6] H. Bean, B.F. Liu, S. Madden, J. Sutton, M.M. Wood, and D.S. Mileti, 2016, Disaster warnings in your pocket: How audiences interpret mobile alerts for an unfamiliar hazard, *Journal of Contingencies and Crisis Management* 24(3):136-147.

[7] While the work done on public response to message length was completed through a DHS research project, the Communications Security, Reliability and Interoperability Council V (CSRIC V) reviewed the technical feasibility of increasing the message length.

to their safety, and to deliver the messages to populations at risk of imminent threats with the goal of maximizing the probability that people take protective actions and minimizing the delay in their taking those actions. Given the extensive body of knowledge built on six decades of research, we have an extensive understanding of key properties of effective alert and warning systems.

This body of knowledge tells us that an effective alert and warning system will be capable of the following:

- Only target and reach people (or their devices)—and people who care about the people and property—at risk from the hazard. Fine-grained geotargeting is important to ensure that those who are not at risk do not receive alerts that do not apply to them. However, others who may not be in the at-risk area may want or need to know about the hazard; the hazard may impact a location of interest. For example, a parent might want to know about hazards that would impact their child's school.
- Communicate impact and recommend protective actions that people can understand and can reasonably take with the guidance provided and tailored to the circumstances of each alert recipient. For example, recommending to shelter in place may not be actionable for mobile home residents; advance recommendations to evacuate or shelter elsewhere would be more helpful.
- Be respected and trusted by the public, emergency managers, other public officials, and the media. Alert originators need to trust that the system will in fact deliver an alert sent in a timely manner and to all planned recipients and the public needs to know that the delivered message is in fact accurate and is from a trusted source. This property will rely on key technical capabilities and system properties of dependability, reliability, resilience, and security.
- Be suitable for all hazards and effective in reaching all at-risk populations. The population impacted by hazards is incredibly diverse in numerous ways, including differences in languages, abilities, and technology access. An alert and warning system needs to support this diversity and communicate to each impacted subpopulation effectively.
- Work well alongside other government and private information sources. As noted above numerous public and private organizations either collect information during a disaster, provide information, or both. Therefore, alert and warning systems must work alongside these services; alerts will need to be easily repurposed for other media or delivery methods).
- Allow for collecting feedback from the alerted population to determine the effectiveness of an alert and give emergency managers better situational awareness during an event. Feedback is needed during a crisis

to immediately understand how the public is responding to the event but is also needed for post hoc analysis so that systems can be improved.

EVOLUTION OF AN INTEGRATED ALERT AND WARNING ECOSYSTEM

The committee envisions an alert and warning system that continually takes advantage of new technologies and reflects the results that emerge from research. In the near term, this will mean increasing adoption of WEA and other existing alert and warning systems, incorporation of current knowledge about public response to craft more effective alert messages, and research focusing on verifying technology capabilities. Existing technologies, such as newer delivery and geotargeting technologies, will need to be adapted for use in alert and warning systems. Long-term, this will involve exploring new technologies, gaining a better understanding of existing technologies, and continued technical, social, and behavioral research to inform the design and operation of future alerting capabilities. These near- and long-term visions for an alerting system underpin the research agenda described in the next section.

Near-Term: Use Current Alert and Warning Technologies and Tools

Near-term goals for an integrated alerting system are twofold: fully adopt and understand current alerting tool and adopt newer technologies for use within that system.

Fully Adopt and Understand Current Alerting Tools

As of August 8, 2016, less than a third of U.S. counties have registered to use the Integrated Public Alert and Warning System[8] gateway, the system that allows message originators to send WEA messages. Only 387 wireless emergency alerts have been originated by state or local governments since WEA came online; by comparison the National Weather Service has sent approximately two million alerts.[9] An increased use of WEA by emergency officials could mean reaching additional populations, and increased use would also improve familiarity with the systems, which could improve public response times.

---

[8] IPAWS was created under Executive Order 13407 to integrate various alerting systems—Emergency Alert System, National Warning System, Wireless Emergency Alerts, and NOAA Weather Radio All Hazards—into one modern network. IPAWS takes advantage of the Common Alerting Protocol, an XML-based data format for exchanging alerts and warnings.

[9] M. Lucero, FEMA IPAWS Division, "IPAWS Evolution," presentation to the committee on August 9, 2016.

Research is also needed to understand the implications of new FCC rules for WEA, which expands the message length to 360 characters and allows the use of Web links (URLs) in messages. Although the new rules will provide new opportunities for emergency managers who have struggled to provide useful information in 90 characters, research is needed to determine what information to include and how to best display additional information, in the WEA message itself and on any media it links to. Furthermore, research is needed to better understand what message lengths are technically feasible and what message length elicits the best public response.

Adopt Newer Technologies for Use Within That System

WEA was developed prior to the wide use of smartphones and newer cellular network technologies. Incorporation of newer technologies could address many shortcomings of WEA, including a host of accessibility, functionality, and other concerns. These advances include the following:

- Modernize delivery technologies. The immediate opportunity to modernize is to switch from 2nd/3rd generation short message service-based (cell broadcast) to (4th generation) long-term evolution (LTE) broadcast as the primary bearer.[10]
- Diversify communications technologies in handsets to help distribute alert messages when cellular network congestion or failure occurs. Short-range communications technology such as Bluetooth and WiFi could be used to forward messages locally while FM radio provides an alternate long-range technology.
- Support the use of location information stored on the handset to improve the precision of geotargeting by determining if a device is located within the targeted area and whether an alert should be displayed. Smart phones, using GPS and other technologies, are very capable of not only knowing where a phone is but also where it has been (and potentially where it is likely to be in the future).
- Adopting a more "app"-like approach to WEA, as opposed to the current simple text like approach. Such an approach would allow for the incorporation of more flexibility so that alert and warning capabilities can be upgraded more easily as understanding of public response and technology capabilities change. For example, the software on smartphones that supports WEA alerts could be moved from the operating system (which on some phones may not be frequently updated) to a more easily updated app distributed through the normal application distribution channels.

---

[10] LTE Broadcast (or multicast) provides faster delivery and supports a larger content size.

- Provide mechanisms for performance monitoring and user feedback to facilitate studies related to perceived relevance (by seeking user feedback and/or inferring action taken), coverage (how many users did and did not receive a message), and message delivery latency.

Long-Term: Incentivize the Building of an Integrated Alert and Warning Ecosystem

The increasing number of connected devices, sensor networks, and mobile phone capabilities provide significant opportunity to detect events, deliver well-vetted alerts over numerous channels, and gather feedback on how these alerts are perceived. As a result, the overall ecosystem will be both rich and complex. A framework can be developed that allows for gathering alerts from multiple sources and making those available for other third-party applications and incentives participation by those third parties. Such a framework could leverage the increasingly advanced capabilities of connected personal devices to support applications that will factor in user preferences and the dynamic context and relationships they find themselves in to present the information in effective ways. The emergence of frameworks like Apple's Homekit and Google Home demonstrate the feasibility of such an approach. A framework could track the relevance, fidelity, veracity, and uncertainty of the data, contributing to building a better system, and provide mechanisms to enable revision of stale or incorrect information. Such a framework could also be designed to decouple the content of messages and data from the channels through which the content is delivered, eliminating the need to create separate stovepiped systems and use all available modes of communication, ranging from managed cellular systems to opportunistic peer-to-peer systems.

While short-term evolution focuses on improvements on or extension of WEA, the committee also foresees a wider capability for IPAWS as a central tool to an integrated alert and warning ecosystem that draws on a wide variety of data sources for better understanding emergencies and the public response and that encompasses a wide range of potential technologies and devices for delivering messages. Box 2.1 explores example scenarios of how advances in technology might be used in an alert and warning ecosystem.

Envisioning such an advanced system requires exploring questions around technical feasibility and implementation and an understanding of how these tools will impact public response. However, social and behavioral research already informs us of properties an ecosystem should have, including the following:

> **BOX 2.1**
> **Envisioning Future Alerting**
>
> The combination of connected devices (or Internet of Things) and sensor data could provide increasing capabilities to not only alert or warn an at-risk population, but also increase situational awareness of emergency responders and managers, improve public response, and even potentially decrease damage by use of automation. Examples of these future capabilities include the following:
>
> - A sensor network detects ground movement associated with an earthquake, triggering an immediate alert to the public within the potential impact area. This alert would also automatically turn off gas valves, stop trains or elevators, or other damage-limiting triggers. (This already exists in some areas.)
> - An evacuation order is sent to various in-vehicle navigation systems and navigation applications. These navigation systems reroute evacuation traffic across diverse routes to prevent traffic delays and also update frequently to route around closed roads.
> - In-building location tools navigate persons to safety during an active shooter scenario and inform responders where at-risk populations are.
> - A personal assistant device (such as Alexa or Google Home) alerts a resident that they are currently under a tornado warning. The resident asks the device to describe the difference between a warning and a watch and follows up with questions of what the best protective action might be.
> - An individual who is differently abled receives protective action that considers the person's capabilities and information on how a caretaker might best respond.
> - A smoke detector alerts not only the residents of a home but also sends an alert to neighboring homes so that they may take necessary actions.
>
> Of course, development of any of these capabilities will first require an integrated privacy and security assessment and the development of an appropriate framework for managing the privacy, trust, and security issues that could arise with such functions.

- Using technologies that are privacy preserving. For example, location and other contextual information can be stored locally on a smartphone, and applications can use this information to decide when and how to display messages.
- Assuring end-to-end service availability and the integrity of valid messages, preventing spoofed messages and spoofed alerts, and assuring system availability from alert origination to message receipt.
- Giving users as much control as possible over what kinds of messages they receive, and alerting control should not be limited to simply on or off.

- Including metadata in alerting systems that can be used in combination with user preference to determine when and how to present alerts.
- Integrating messages across communication channels, given the wide number of available technologies. For example, IPAWS messages could be made available as a data stream for private industry to use freely in weather applications, navigation systems, social media streams, and the like.
- Making alerting systems devices agnostic and able to support more than one modality of information presentation. For example, both text and voice alerts can be provided on mobile devices.
- Reflecting a better understanding of the information needs of emergency managers to quickly analyze data generated via social media.
- Using IoT devices and other embedded sensors to detect, analyze, and categorize potential events, send alerts, and potentially automate certain protective actions for minimizing potential damages.
- Incorporating available communications technologies, such as mesh networking and FM broadcast signals,[11] to increase the ability to deliver information in the event that primary communication networks fail.
- Adapting message content and format to the context and needs of the end-user, for example, considering location of device, known home location of device owner, language of device owner, disability status, and other context (as selected or entered by the user).

---

[11] Many smartphones have FM radio receiver hardware built into them. There is potential for these to be used to provide information if a cellular network is not functioning or data access is limited for other reasons; however, enabling this function requires the consideration of a number of technical and business issues.

3

A Research Agenda

As Wireless Emergency Alerts (WEAs) and other new technologies have been deployed, research investments, in large part supported by the Department of Homeland Security (DHS), have provided some insight into questions around the use of new technologies for alerting (Appendix B includes summaries of this work). However, to reach the above-envisioned alert and warning system, additional research questions will need to be answered about the use and design of WEA as well as of evolving systems. Given that alerts and warning are inherently interdisciplinary, both a social science phenomenon (their goal is to change public behavior) and a technical phenomenon (technology is required for their assessment and dissemination) this research agenda includes a wide range of sociotechnical questions and highlights the need for social and behavioral scientists and technologists to interact frequently with each other. The agenda is divided into key sections: public response, feedback and monitoring, and technical-capabilities and their impact.

PUBLIC RESPONSE TO ALERTS AND WARNINGS

Ultimately, alert and warning systems need to be designed to elicit the most life- and safety-protecting response from the public. Research has evolved over the last several decades so that we have much more information about how individuals respond to alerts and warnings. Nevertheless, as technologies shift, so do public responses; therefore, continued research is invaluable. These responses rely on several things, including characteristics of the messages themselves, demographics of the individuals who receive the messages, and, given our increased ability to geo-target messages, the understanding of an individual's risk in relationship to their location and the hazard. Research is also needed to understand how best to educate the public about both alerting systems and impacts of and appropriate response to particular hazards.

Message Characteristics

Protective Guidance in Enhanced Media Links

Prior research called for including uniform resource locators (URLs) in WEAs to provide more complete information on the hazard and recommended protective guidance.[1] Practitioners also spoke to the importance of enhanced warnings. For example, Christopher McIntosh, former Virginia statewide interoperable communications coordinator and current director for national government industries at the geographic information system firm Esri, observed that "without context, alerts are just noise."[2]

At this point, it is unclear what information is best included in a WEA message and what information is best included in linked content. Prior DHS-funded research did not examine exactly 360-character messages because the research was conducted before the Federal Communications Commission (FCC) rulemaking that extended WEAs from 90 to 360 characters.[3] Furthermore, concerns remain as network congestion could be caused by people accessing an included link within seconds of receiving a WEA that includes a URL.[4] Alternatively, some research found that message recipients are unlikely to open linked content and that instead they read only a few words of WEA-like messages due to stress responses.[5] Therefore, research is needed to understand what message content should be included in linked media. Also unknown is how to craft WEA mes-

---

[1] M. Wood, H. Bean, B. Liu, and M. Boyd, 2015, *Comprehensive Testing of Imminent Threat Public Messages for Mobile Devices: Final Report*, College Park, MD: National Consortium for the Study of Terrorism and Responses to Terrorism.

[2] C. McIntosh, Esri, presentation to the committee on January 26, 2017.

[3] Federal Communications Commission, "FCC Strengthens Wireless Emergency Alerts as a Public Safety Tool," release date September 29, 2016, https://apps.fcc.gov/edocs_public/attachmatch/DOC-341504A1.pdf.

[4] Federal Communications Commission, "Improving Wireless Emergency Alerts and community-initiated alerting," release date November 19, 2015, https://apps.fcc.gov/edocs_public/attachmatch/FCC-15-154A1.pdf

[5] D. Glik, K. Harrison, M. Davoudi, and D. Riopelle, 2004, Public perceptions and risk communication for botulism, *Biosecurity and Bioterrorism: Biodefense Strategy, Practice, and Science* 2(3):216-223.

sages so that they galvanize people to read the entire message, including potentially life-saving linked content. Research is also needed on whether it is more effective to enhance protective action guidance by using longer, 360-character alerts or by adding links to additional information. A consistent finding across WEA research is that the American public needs education on what the WEA service is as well as what protective actions to take during a variety of hazards.

Expressing Time Until Hazard Impact

The Study of Terrorism and Response to Terrorism (START) research team found that the standard WEA message elements of *guidance* (what to do and how to do it) and *time until impact* (how much time people have to take the recommended action) play major roles relative to other message elements in the outcomes of public understanding and belief of the protective action recommendation and the ability to decide how to respond.[6] The other WEA message content elements are hazard, location, and source. Importantly, the START research team found that WEAs should express time as how much time until impact rather than when the message expires, as is the current practice for WEAs.

WEA messages are designed to alert about imminent threats, which the START research team characterized as hazards occurring within one hour. Other research has extensively examined the optimal timing of warnings. For example, research on tornados finds that the optimal lead time for issuing a tornado warning is from 15 minutes to just over 30 minutes.[7] If too much lead time is provided, people are less likely to follow the protective guidance in a timely manner. Therefore, it is important to understand how to best express lead time.

Opt-In/Opt-Out

Current WEA guidelines allow for opting out of all categories of alerts except for those issued by the U.S. President (which have never been issued). However, individuals now receive messages from an increasing number of sources and delivery channels. Past research suggestions that

---

[6] M. Wood, H. Bean, B. Liu, and M. Boyd, 2015, *Comprehensive Testing of Imminent Threat Public Messages for Mobile Devices: Final Report*, College Park, MD: National Consortium for the Study of Terrorism and Responses to Terrorism.

[7] S. Hoekstra, R. Butterworth, K. Klockow, D.J. Drotzge, and S. Erickson, 2011, A social perspective of warn on forecast: Ideal tornado warning lead time and the general public's perceptions of weather risks, *Weather, Climate & Society* 3(1):128-140; and K.M. Simmons, and D. Sutter, 2009, False alarms, tornado warnings, and tornado casualties, *Weather, Climate & Society* 1(1):38–53.

alerts and warnings should be sent through as many channels as possible, but new research is needed to explore what drives opt-in and opt-out behaviors on WEA and as well as on various platforms, such as third-party applications, or local text alerting systems. While past research[8] supported the delivery of alert messages across as many channels as possible, it is unknown if the increasing number of alerting channels provides the same benefits or if it instead creates a situation of over-alerting, which may result in increased opt-out rates. Furthermore, once we understand optimal times and way to alert an individual, what are technical solutions to avoid over-alerting?

Message Length

While we know a lot about what a message should contain, less is known about how to best present this information. Recent research has provided clear evidence that message length influences response; messages that can fit in the initial 90-character length of a WEA message and the 140 characters of Twitter foster milling[9] behavior and delayed response.[10] At this point, it is unclear what information is best included in a 360-character WEA message and what information is best included in linked content. Prior WEA research did not examine 360-character messages because the research was conducted before the pending FCC rulemaking that extended WEAs from 90 to 360 characters.[11] However, research suggests that a mes-

---

[8] Considering the sharing of emergency alerting messages, there are several benefits to an ecosystem that incorporates multiple different channels. One is redundancy—i.e., a message distributed via different channels (via different technological infrastructures) is more likely to reach its intended recipients in cases where some infrastructures are disrupted. The second benefit is diversity—i.e., people rely on different types of media platforms (e.g. due to cultural preference, education, or accessibility), and so messages spread across diverse channels will reach a greater number of people. A third benefit is that people are more likely to accept and respond to emergency messages when they receive them from multiple channels and in different formats. This latter benefit, which was identified in classic studies, requires more research to confirm and better understand in the context of this new media ecosystem.

[9] "Milling" refers to the process in which people seek to confirm an alert or warning, a process that has been observed across all hazard types, warning delivery technology, or message sources.

[10] H. Bean, M. Wood, D. Mileti, B.F. Liu, J. Sutton, and S. Madden, 2013, *Phase II Interim Report on Results from Experiments, Think-out-Loud, and Focus Groups, Comprehensive Testing of Messages for Mobile Devices*, Report to the Homeland Security Advanced Research Projects Agency, Science and Technology Directorate, U.S. Department of Homeland Security, College Park, MD: National Consortium for the Study of Terrorism and Responses to Terrorism, University of Maryland.

[11] Federal Communications Commission, "FCC Strengthens Wireless Emergency Alerts as a Public Safety Tool," release date September 29, 2016, https://apps.fcc.gov/edocs_public/attachmatch/DOC-341504A1.pdf.

sage length of 1,380, the maximum number of characters supported by the Common Alerting Protocol (CAP) standards additional information field, does reduce response time.[12] Research is needed to understand public response to messages that fit into 360 characters and, given that optimal message length is relevant to any text-based alerting system, continued research should be done to understand the optimal minimum length that can elicit the appropriate protective action from an alerted population.

Demographics

Language and Dialect

WEA currently supports messages in Spanish but this capability falls well short of the linguistic diversity of the U.S. population. What technical challenges exist in transmitting multiple languages, or relying on the receiving device to translate messages? Additionally, protective action language, such as "shelter in place," might be challenging to translate to various languages and dialects. Research is needed to understand the limitations of language translations (in particular machine translations)—and determine what constitutes "good enough" language so that message templates can be automatically translated—for differing languages and dialects.

Adapting to Differing Abilities

Mobile devices exist that can be used by a wide set of differently abled individuals, these include a range of tools, such as use of vibration cadences to Braille phones and text-to-speech and speech-to-text tools based on needs of physically challenged individuals. Research is needed to understand best way for enabling specific customization (translating and delivering) alerts and warnings to physically and cognitively challenged individuals. What other technologies exist to support information dissemination to differently abled individuals? How can protective action instructions shift to support diverse populations—including those of differing ages and abilities—and their caregivers? Questions around literacy are also important, in terms of both age (given that children under 10 may receive an alert on their cell phone) and reading comprehension for older adults.

---

[12] H. Bean, B.F. Liu, S. Madden, J. Sutton, M.M. Wood, and D.S. Mileti, 2016, Disaster warnings in your pocket: How audiences interpret mobile alerts for an unfamiliar hazard, *Journal of Contingencies and Crisis Management* 24(3):136-147.

Technology Access

While a large section of the population uses smartphones, there are still others who choose not to use smartphones or use them sporadically. Considering the diversity in communication habits and availability of technology, alert and warning systems will need to consider various technologies to reach at-risk populations.

Geotargeting Alerts and Warnings

Communicating Location

Research suggests that people do not easily understand messages that contain a map reflecting the at-risk area.[13] We know that a map that shows the risk area alone is not useful. In fact, it can be counterproductive. What is needed is research on how to best communicate, possibly through visualizations, about the location of the message receiver versus the area of impact. Research is needed to determine the best way to graphically display that an individual is in an at-risk location.

Determining Locations of Interest

Individuals want to be alerted not only when they are at risk but also, for example, when their children may be at risk at school or their home may be at risk. These locations of interest can be difficult to determine. Most systems rely on the receiving device being currently within the designated warning area; if a person works outside of a WEA alert area but their home is within the WEA alert area, they will not receive the message that there home is at risk. Most receive these alerts via subscription services, such as those provided by a school system or county. Are there technical solutions so that locations of interest can be dynamically updated (rather than manually updated by the end-user)?

Location-Based Protective Action

The best protective action for an individual may vary across the impacted area—shelter in place versus evacuation. Furthermore, individuals could be prescribed specific evacuation routes to spread traffic over different routes. What are the technical challenges to these technologies? Applications such as Waze could provide some of these capabilities. What

---

[13] M. Wood, H. Bean, B. Liu, and M. Boyd, 2015, *Comprehensive Testing of Imminent Threat Public Messages for Mobile Devices: Final Report*, College Park, MD: National Consortium for the Study of Terrorism and Responses to Terrorism.

are limitations to implementing these for disaster responses? How might we encourage use of these tools?

In-Building Location

Indoor location capabilities are already being deployed in some areas, chiefly for marketing reasons as well as for meeting wireless E911 location requirements. The FCC now requires that carriers provide location information to within 50 meters of a caller's location (inside or outside of buildings) for 40 percent of the cases currently and in the near future for 60 percent of the cases. This requirement includes both horizontal and vertical location information. However, determining elevation (which floor) is challenging and is an important research topic. Knowledge of a person's location within a building could be used to determine the best evacuation route or if the individual should instead shelter in place.

Community Engagement

New tools and technologies support communications between members of a community; for example, NextDoor allows people to quickly identify neighbors and communicate with those people who reside either in their neighborhood or nearby. NextDoor is already being used by public safety organizations to educate the public;[14] however, little is known about how these tools are used during hazards. As discussed in Chapter 1, collaborative tools have been used to provide assistance during a disaster but mostly fueled by volunteers outside of the area. Less is known about how local residents interact with each other to build resilience and educate others about particular hazards and how that information propagates throughout a social group. This issue of course intersects with other research areas around reposting information on social media and disaster education, but is itself an important area of research.

Disaster and Alerting Education

Across the DHS-funded studies, research participants were found to be unfamiliar with WEAs[15] despite the Ad Council having partnered

---

[14] M. Helft, A Facebook for crime fighters, *Fortune.com*, July 1, 2014, http://fortune.com/2014/07/01/nextdoor-local-neighborhood-social-network-police/.

[15] M. Wood, H. Bean, B. Liu, and M. Boyd, 2015, *Comprehensive Testing of Imminent Threat Public Messages for Mobile Devices: Final Report*, College Park, MD: National Consortium for the Study of Terrorism and Responses to Terrorism; B. Kar, University of Southern Mississippi, "An Integrated Approach to Geo-Target At-Risk Communities and Deploy Effective Crisis Communication Approaches," presentation to the committee on September 1, 2016;

A RESEARCH AGENDA 63

with DHS to promote the WEA service.[16] Given the character constraints of WEAs, researchers found that participants required more information to properly execute recommended protective actions.[17]

Very limited research exists to determine what makes for effective disaster public education. Research finds that current public education campaigns typically are ineffective because they are not specific enough and do not contain content that motivates behavior change.[18] More research is needed to determine how to motivate behavior change as well as what other factors contribute to successful public disaster education campaigns.

In terms of in-person training, research points to positive results from these public education initiatives. For example, in one study researchers found that, after receiving instruction on relevant meteorological principles, participants successfully applied their new knowledge to make risk inferences from hazard graphics.[19] As another example, an evaluation of research on disaster education programs concluded that these programs are effective at increasing children's disaster knowledge and preparedness as well as household preparedness.[20]

Finally, emergency managers have just begun to integrate gamification into public education, and it is too soon to tell how effective this approach is for increasing individual, family, and community disaster preparedness. For example, in 2013 the Centers for Disease Control and Prevention launched the "Solve the Outbreak" mobile app, which allows users to be "disease detectives" through obtaining clues, analyzing data, solving scenarios, and saving lives in the game. So far, the app has been downloaded more than 12,000 times. Research on whether this app or

---

D. Glik, University of California, Los Angeles, "WEA Messages: Impact on Physiological, Emotional, Cognitive and Behavioral Responses," presentation to committee on September 1, 2016; and others.

[16] Ad Council, "Emergency Preparedness – Wireless Alerts," https://www.adcouncil.org/Our-Campaigns/Safety/Emergency-Preparedness-Wireless-Alerts, accessed August 22, 2017.

[17] M. Wood, H. Bean, B. Liu, and M. Boyd, 2015, *Comprehensive Testing of Imminent Threat Public Messages for Mobile Devices: Final Report*, College Park, MD: National Consortium for the Study of Terrorism and Responses to Terrorism.

[18] B.J. Adame and C.H. Miller, 2015, Vested interest, disaster preparedness, and strategic campaign message design, *Health Communication* 30(3):271-281; J.D. Fraustino and L. Ma, 2015, CDC's use of social media and humor in a risk campaign – Preparedness 101: Zombie apocalypse, *Journal of Applied Communication Research* 43(2):222-241; and M.M. Turner and J.C. Underhill, 2012, Motivating emergency preparedness behaviors: The differential effects of guild appeals and actually anticipating guilty feelings, *Communication Quarterly* 60(4):545-559.

[19] M. Canham and M. Hegarty, 2010, Effects of knowledge and display design on comprehension of complex graphics, *Learning and Instruction* 20(2):155-166.

[20] V.A. Johnson, K.R. Ronan, D.M. Johnston, and R. Peace, 2014, Evaluations of disaster education programs for children: A methodological review, *International Journal of Disaster Reduction* 9(1):107-123.

others improve preparedness and capacity to effectively respond to warnings during disasters remains to be documented. A similar research topic is how best to use tools, such as video and animation, to model protective actions and the use of tools that provide education as a disaster unfolds.

POST-ALERT FEEDBACK AND MONITORING FOR EMERGENCY ORGANIZATIONS

Technology is needed that solicits feedback from message recipients to help understand how the public is responding to messages and what additional information might be needed. While some tools exist to help individuals in the emergency operations center to harvest information from social media and potential feedback mechanisms in alerting applications on mobile devices, these tools will need to be more readily available. Perhaps more importantly, research is needed to understand what information would be most helpful to emergency managers and social science researchers and how best to collect the information. Not only would these additional data help emergency managers during disasters, but they could also serve to validate laboratory experiments. Several researchers have conducted experiments to explore word choice, message content, and character length. While these experiments provide valuable data, real-world analysis could provide validation and further information on public response. Tools, including those that employ machine learning and other artificial intelligence techniques, are also needed that can quickly understand and process collected information.

Consistent, well-understood, and insightful measurements can inform (and improve) response to future hazards (Box 3.1 lists information that could be valuable). Such a data-driven experimental framework would be of great interest to multiple stakeholders, including emergency managers, researchers, technologists, and so on. By building measurement into the alerts and warning system itself, researchers can gain supporting evidence for findings made in lab studies (e.g., what is the optimal message length? should we include a map or not?). By sharing information across hazards, we can learn from past experiences to create new best practices. Feedback during the life cycle of a hazard can also be integrated into future responses within the same incident. For example, low response rates to an initial message could lead to more aggressive message content in a follow-on message.

In parallel with measurements of the messages themselves, we also encourage new measurements of ancillary supporting technologies. For example, what level of engagement is seen on social media and local news websites? What content was most engaged with? What fraction of users in a region used Waze? And so on. We anticipate new data-sharing initiatives for the multiple stakeholders in the alerts and warnings ecosystem.

> **BOX 3.1**
> **Measuring the Effectiveness of Alerts and Delivery Mechanisms**
>
> Performance monitoring can be undertaken throughout the life cycle of an alert. In some cases, new standards may need to be adopted or new technologies created to support monitoring and feedback. Similarly, new social science research will be needed to assess the usefulness of metrics. Examples of possible measurements of interest could include the following:
>
> - *Coverage.* How many people (or devices) received a message? And what fraction of all affected people is this? What are the characteristics of those who did not receive a message, and how can this gap be closed?
> - *Message engagement.* How long did people view a message? Was it immediately dismissed? Tracking other activity on a smartphone may gain additional evidence of engagement with a message and follow-up activities (e.g., phoning a friend, checking social media feeds). Of course, privacy is a paramount concern here.
> - *Actions taken.* Upon receiving a message, how many people take action? And what actions are these? Perhaps structured responses can be created so that people can do a "safety check" or equivalent, depending on the nature of the alert. Careful design of feedback mechanisms is important.
> - *Latency.* What is the time from the inciting incident to when the message was received by users? Latency here could be measured at different levels of granularity—including network performance measures based on when a message is injected, time from inciting incident to message injection itself, and so on.
> - *Translation effectiveness.* Measurements of engagement with a message can also provide downstream understanding of the quality of upstream decisions. For example, machine translation of messages may achieve a particular "accuracy" in off-line assessments, but through measurements of these translations in practice we can provide supporting evidence of the quality of the translation (e.g., an off-line translation engine may have 98 percent accuracy, but lead to much lower engagement once issued).

For example, it could be beneficial for social media companies to provide aggregate (anonymized) measurements in the aftermath of an alert.

TECHNICAL CHALLENGES AND THEIR IMPACT

Delivery Technology

Enhanced Cell Broadcast and Network Capabilities

Cellular networks have become the most popular access mechanism in the current scenario due to the wide use of the device. However, there are limitations to the current implementation of cell broadcast—limitations in coverage, capability of handling message size, and inability to facilitate two-way communication. Research is needed to understand capabilities and features that may take advantage of newer broadcast technologies and supplement older technologies. Research questions include the following: Is it feasible to combine multiple cell broadcast messages core into a single, longer alert message?[21] What will the impact of including a URL be on network capacity? Recent WEA testing has indicated that messages are not always delivered as expected;[22] what is creating these errors? As we move to newer wireless network standards, is it possible to flexibly select among 90-, 360-, and 720-character-long messages to best make use of existing 2.5G to 3G networks as well as new upcoming 4G-LTE (long-term evolution) networks? (See Box 3.2.)

Multimodal Transport of Emergency Alerts

Today WEA is designed only for cellular transport. However, cell phones can receive data through a variety of wireless communications standards. For example, phones are commonly connected through home or public WiFi hotspots. How can WEA be adapted so that it can use multiple channels to increase the likelihood of successful delivery to the end-user? How can a single message be delivered through an increasing number of delivery channels—including government and private channels?

Bypassing Network Failure

During hazards, some cellular networks may not function properly, owing either to overload or to infrastructure damage.[23] One existing alternative is the NOAA Weather Radio All Hazards, which relies on a national network of dedicated transmitters and users who obtain suitably equipped receivers.[24] Several other technologies exist that might support message receipt, including mesh networks, peer-to-peer communication,

---

[21] The Alliance for Telecommunications Industry Solutions completed feasibility studies on message length in 2015 and recommended the expansion to the current 360-character limit for next-generation networks (https://access.atis.org/apps/group_public/download.php/25045/ATIS-0700023.pdf). Additional feasibility studies may be necessary to understand future limitations.

[22] Federal Communications Commission, 2017, "Report: September 28, 2016 Nationwide EAS Test," https://apps.fcc.gov/edocs_public/attachmatch/DOC-344518A1.pdf, and in briefing and discussions with the committee.

[23] The National Research Council reviewed the impact of network usage during disasters in its report *The Internet Under Crisis Conditions: Learning from September 11* (Washington, DC: The National Academies Press).

[24] This text was modified after prepublication.

> **BOX 3.2**
> **The Role of Standards in Delivering Alerts and Warnings**
>
> Technical standards play an important role in communication systems. As such, a single emergency alert message standard is needed to communicate alerts over a variety of communications infrastructure (e.g., broadcast TV and radio, cellular, and wired media), thereby reaching the end user over variety of last-mile technologies (e.g., cellular, cable, DSL, WiFi, Bluetooth). The Common Alerting Protocol (CAP) attempts to serve this purpose.
>
> CAP, which is used by IPAWS, provides a standardized data profile that defines a consistent way for alert and warning messages to be distributed among the involved entities (alert originators, the Federal Emergency Management Agency, the National Oceanic and Atmospheric Administration, public safety, broadcasters and mobile operators. This standard takes the form of an XML format and is one of the many standardized XML formats that the Organization for the Advancement of Structured Information Standards maintains. The CAP standard describes how the information is formatted. While CAP provides a common format, any changes in the type of information exchanged in CAP or any changes in how the carriers might communicate this information to a wireless subscriber requires additional standards and versions of those standards.
>
> Standards efforts can take years to complete even relatively simple changes to existing standards. For example, it took several years for the cellular standard body (3GPP) to recently come up with a specification that allows the transmission of 360-character Wireless Emergency Alerts messages. Alert and warning standards—and communication standards that support delivery of alerts and warnings—could evolve more rapidly to incorporate sociotechnical understanding of public response.

and FM radio transmission. For example, with peer-to-peer communication techniques, such as those used by FireChat, messages might be relayed to people lacking a direct network connection. Research is needed to validate the efficacy of these various technologies and understand the implementation of these tools.

Battery Life Management on End User Devices

Power resources can be quickly drained during disasters as users attempt to find information. If main power to a home is out, a user will also lose access to Internet service and forced to rely on cellular data service that uses more battery power than in-home wireless, and will be unable to recharge devices. Research is needed to determine possibilities for reserving power resources to support alert and warning delivery, implementing battery preservation technologies across available platforms, and possibly providing information back to first responders about an inability to receive messages owing to power concerns.

Role of Connected Devices

As the Internet of Things (IoT) grows, more devices in homes and throughout the environment will be available not only as an alerting channel but also to detect emergencies and potential risks. To make most effective use of these opportunities, several questions will need to be explored.

Aggregating Data

If each home were to be outfitted with a mini-weather station or stream gauges were to be deployed pervasively, what data would be most helpful and how can data be trusted enough to automatically issue an alert and warning? Automation of some alerting, based on aggregate data, would resolve latency issues around fast-moving hazards. For example, in several areas of Japan, earthquakes are detected and elevators, trains, and gas valves are immediately cut off. Could similar automation be used for other events such as an active shooter event—gunfire on campus is detected and classroom doors are locked? Machine learning is needed to understand how systems can become sufficiently data intensive and have enough situational awareness to suggest the best protective action to take—i.e., such as when to shelter in place, when to evacuate, and where to evacuate.

Best Devices for Alerting

If most electronics can deliver some sort of information to users, which devices should be used to issue which hazards? Could the increasing number of smart devices better communicate appropriate protective action? For example, if a disaster results in a boil water order, could a smart refrigerator present an alert when the water dispenser is used? (See Box 3.3.)

Milling with Virtual Assistants

As more homes are equipped with virtual assistants, such as in an Alexa or Google Home, what role will they play in milling behavior? One can envision receiving an alert from Alexa that a tornado warning has been issued and a user has asked Alexa to explain the difference between a tornado watch and a tornado warning or is asking what the best protective action might be. This might on the one hand reduce milling times because confirming information can be obtained quickly and might on

> **BOX 3.3**
> **In-Vehicle Alerting**
>
> Given the ubiquity of terrestrial radio in vehicles, drivers historically received alerts and warnings through the Emergency Alert System. With the deployment of Wireless Emergency Alerts and the increasing number of weather applications that support alerting, the mobile phone is also increasingly the source of alerts for drivers. Navigational tools, both those built into the vehicle and mobile applications, can also provide alerts, although they tend, naturally, to focus on traffic hazards. How can these tools be adopted to provide better alerts and warnings, suggestions for protective action, and situational awareness back to emergency responders?
>
> The navigation service Waze has worked with local emergency managers and other volunteers to provide some hazard-alerting capabilities, including the following:
>
> - Emergency shelter locations along with Web and phone links
> - Information about the status of gas station
> - Information about unusual traffic patterns
> - Real-time updates on road closures and other hazard conditions[1]
>
> For example, drivers—and in the future, autonomous vehicles—can be automatically rerouted around flooded roadways or overcrowded evacuation routes.
>
> ---
> [1] Adapted from Google, "Crisis Event Support on Waze," https://support.google.com/waze/partners/answer/6342326?hl=en, accessed September 7, 2017.

the other hand extend milling because it introduces new avenues for extended information seeking.

Security, Trust, and Privacy

A system that instructs large populations to take a particular action may represent a significant target for attacks on service availability, compromises of the integrity of valid messages, and spoofed messages. Emergency alerting systems have been directly compromised already, including the use of false Emergency Alert System (EAS) tones, resulting in the issuance of false alerts on radio and TV stations in several states in 2014.[25] (Box 3.4 lists some breaches that have already occurred.) Indeed, a 2014

---
[25] R. Wimberly, "$1 Million Fine for Misusing the Emergency Alert System," release date May 19, 2015, http://www.govtech.com/em/emergency-blogs/alerts/1000000-Fine-for-MisUsing-Emergency-Alert-System.html.

> **BOX 3.4**
> **Alert System Breaches**
>
> In November 2010, Iowa's alerting system was compromised, resulting in the issuance of a false AMBER Alert.[1]
> On February 11, 2013, a hacker sent out an emergency alert that read "dead bodies are rising from their graves," in some counties in Great Falls, Montana.[2] In a similar case, on February 12, 2013, two Michigan Television Stations began displaying a fake alert message warning people of zombie attacks in various Michigan counties after the emergency alert system was hacked.[3]
> On September 28, 2016, FEMA's Emergency Alert System was hacked and television viewers in Utica, New York, received a warning on their screen of a pending hazardous materials disaster somewhere in the United States.[4]
> On February 8, 2017, a hacker set off all 156 emergency sirens in Dallas, Texas.[5]
>
> ---
>
> [1] Amber Alert, 2011, Hacked: Lessons from the attack on the Iowa AMBER Alert System, *The Amber Advocate* 5(2):7.
> [2] D. Moye, "KRTV's Emergency Alert System Hacked to Warn of Fake Zombie Apocalypse (VIDEO)," last update February 16, 2013, http://www.huffingtonpost.com/2013/02/11/krtv-fake-zombie-alert_n_2665469.html.
> [3] *Huffington Post*, "Zombie Warning Shown on Michigan TV Stations After Emergency Alert Systems Hacked (VIDEO)," release date February 12, 2013, http://www.huffingtonpost.com/2013/02/12/zombie-warning-michigan-tv-alert-video_n_2671044.html.
> [4] Off the Grid, "FEMA's Emergency Alert System Hacked: Warns of Hazardous Materials Disaster," release date September 29, 2016, https://offgridsurvival.com/femas-emergency-alert-system-hacked-warns-hazardous-materials-disaster/.
> [5] D. Stanglin, 2017, "Hacker Sets Off All 156 Emergency Sirens in Dallas," *USA Today*.

report from Carnegie Mellon University's Software Engineering Institute[26] identifies the following attack vectors that specifically target WEA:

- An outsider obtaining the proper credentials to send malicious CAP-compliant messages, for example, to direct people toward a dangerous location rather than away from it.
- Malicious code that has infected an alert origination service could prevent an operator from posting a new alert, and hence delaying notification.
- An insider may spoof a colleague's identity to send an illegitimate CAP-compliant message, thereby spreading false information and undermining the colleague's reputation.

---

[26] Carnegie Mellon University, 2014, *Wireless Emergency Alerts (WEA) Cybersecurity Risk Management Strategy for Alert Originators*, CMU/SE I-2013-SR-018, Pittsburgh, PA.

- Malicious attacks could make communication channels unavailable, so that an operator could not distribute an alert message through the Integrated Public Alert and Warning System.

These and related threats highlight the critical cybersecurity challenges with maintaining the integrity of existing and future alert and warning systems.

Alerts that can be reliably and undeniably attributed to their actual author is essential to warning system use. This is a particular problem when warning credentials are issued to a jurisdiction or agency rather than to the individual that should be held accountable. If the only feasible sanction for a misuse of a warning system is to threaten to cut off warning system access to an entire jurisdiction, such a sanction is unlikely to be perceived as a genuine threat. Barriers to providing personal credentials need to be studied and remedies devised. Furthermore, research on password management and security could be incorporated into system training tools.

Spoofing, in particular, has been recognized as a threat to the validity of many communication channels, including email, the web, GPS data,[27] and sensor data, among many others. Already, we have seen spoofing attacks on social media to post fake messages as in the April 2013 case of hackers taking control of the official Associated Press (AP) Twitter account to post false reports about an explosion at the White House. Though discovered and corrected within minutes, the spoofed post led to a 100-point drop in the Dow Jones Industrial Average. While the drop was only momentary—the unsophisticated tweet did not follow AP style and lacked other indicators of legitimacy—more sophisticated spoofs in the future could attack multiple channels at once (e.g., the AP Twitter account, Facebook presence, and their wire service) while adopting more credible language to create even more uncertainty.

Furthermore, as the system takes advantage of these large data sets and harnessing information from the public, misinformation can pose a challenge. A misunderstanding by the public and poor reporting can create misinformation, but it can also be inserted intentionally. Quickly detecting and correcting poor information will be a valuable system capability. (See Box 3.5.)

As emergency managers begin harnessing information from users and social media and provide geographically relevant information, concerns around user privacy arise. How can we take advantage of these tools while still protecting end-user privacy?

---

[27] N.O. Tippenhauer, C. Pöpper, K.B. Rasmussen, and S. Čapkun, 2011, On the requirements for successful GPS spoofing attacks, pp. 75-86 in *CCS '11 Proceedings of the 18th ACM Conference on Computer and Communications Security*; https://dl.acm.org.

BOX 3.5
Using Social Media to Detect and Address Rumors and Misinformation

Rumors and misinformation have been identified as a significant threat to the utility of social media and other online tools during disaster events. Many emergency responders cite misinformation on social media as a major concern and a barrier to adopting social media in their work.[1]

However, rumors in the emergency response context are not new and not specific to social media or the Internet. Rumors have always been a feature of the human response to disaster as people come together to try to make sense of an imperfect information space under conditions on high anxiety and high uncertainty. Indeed, though rumoring is often stigmatized and associated with the spread of misinformation, the act of rumoring may serve an important purpose for a community trying to come to terms with a disruptive event. Within this view, rumors can be false or they can turn out to be true or partially true.

This conceptualization differentiates between "rumoring" and the intentional spread of misinformation or disinformation. However, the latter types are also an important part of the information space surrounding disaster events. *Misinformation* can be defined as factually false information that spreads accidentally or intentionally for non-malicious purposes. We see this on social media, for example in repurposed photos (e.g., sharks swimming through flood waters) that are spread with the motivation, perhaps, of gaining visibility and followers for a social media account. *Disinformation* includes false or misleading information purposefully spread to mislead or confuse others—possibly for malicious purposes.[2] Disinformation represents a particularly insidious threat, especially in connected information environments where malicious actors from outside of the affected area may try to interfere with response efforts or intentionally manipulate behavior to put people in harm's way.

Though rumors (and misinformation and disinformation) existed before the Internet, online systems and especially social media have altered the scale of rumoring after disaster events. More people participate—including people from places that are remote from the affected area. Social media features such as the retweet button on Twitter or the share button on Facebook allow people to quickly and easily pass along information—often before they check on the validity of the information or the credibility of its source.

These online sharing practices generate information cascades, as information (including rumors) quickly spreads through these large, interconnected, global networks.

Due to the de-contextualization of online content—i.e., the loss of a clear connection to its original source, including when and where it was published and by whom—it can be difficult to assess the credibility of this kind of information as it spreads.

Further complicating this issue, research suggests that posts that promote false rumors are much more likely to be spread than posts that correct false rumors.[3] Additionally, after rumors are corrected and/or fade away, they occasionally resurface (sometimes years later) and begin to spread anew. This means that, despite early praise of the "self-correcting crowd," online platforms are much better at facilitating the spread of rumors than at correcting or debunking them.

Research suggests that emergency responders who choose to engage online do see the correction of rumors and misinformation as a component of their job.[4] Many spend time monitoring social media for rumors and challenging and/or correcting them. Though there have been arguments that citizens are either unable to identify credible messages from authorities and/or that citizens have active distrust of some official response organizations, preliminary research has demonstrated that rapid corrections to online rumors by official accounts can help to dampen and even stop the propagation of rumors.[5] However, practices for correcting rumors are still dynamic and largely improvised. More research is needed to determine the best strategies—e.g., who should correct, how, and when.

Considering the goal here of informing emergency alerting practices, responders should be encouraged to monitor social media and other online sources to identify and address rumors, misinformation and disinformation that could impact the affected community, and response efforts within that community. If community members are receiving their information from these channels, then responders and other information mediators should assume an active role in curating this content, and research provides some evidence to support the ability for official responders to play a productive role. Responders should also be encouraged to develop protocols for using social media and other outlets to address rumors, misinformation, and disinformation as quickly as possible, though with intentionality that aligns with community norms and expectations within these environments. As those norms and expectations are still evolving, it is important to continue gathering data about and conducting research on the use of these platforms during emergency events.

Recently, discussions about the presence and implications of "fake news" have sparked conversations about educational interventions to help people become better consumers of online content. Considering the emergency response context, there may be opportunity to do educational outreach to help people identify credible and trustworthy sources both before and during emergency events. Research is needed to identify the best approaches for this kind of intervention.

---

[1] L. Plotnick and S.R. Hiltz, 2016, Barriers to use of social media by emergency managers, *Journal of Homeland Security and Emergency Management* 13(2):247-277.

[2] C.B. Schenk and D.C. Sicker, 2011, Finding Event-Specific Influencers in Dynamic Social Networks, *2011 IEEE Third International Conference on Privacy, Security, Risk and Trust and 2011 IEEE Third International Conference on Social Computing*, doi: 10.1109/PASSAT/SocialComm.2011.100.

[3] K. Starbird, J. Maddock, M. Orand, P. Achterman, and R.M. Mason, 2014, Rumors, False Flags, and Digital Vigilantes: Misinformation on Twitter after the 2013 Boston Marathon Bombing, pp. 654-662 in *iConference 2014 Proceedings*, http://hdl.handle.net/2142/47417.

[4] A.L. Hughes and L. Palen, 2012, The evolving role of the public information officer: An examination of social media in emergency management, *Journal of Homeland Security and Emergency Management* 9(1):1-20.

[5] C. Andrews, E. Fichet, Y. Ding, E.S. Spiro, and K. Starbird, 2016, Keeping up with the Tweet-Dashians: The impact of 'official' accounts on online rumoring, pp. 452-456 in *Proceedings of the 19th ACM Conference on Computer-Supported Cooperative Work & Social Computing (CSCW 2016)*, https//dl.acm.org.

4

Challenges to Building Better Alerting Systems

Several challenges, outside of the needed research, exist in building a better alert and warning system. Obviously, alerting and warnings around weather are only as good as the forecasting tools that predict impending severe weather events.

ADOPTION OF ALERT AND WARNING SYSTEMS

The lack of adoption by alert originators (AOs) of the full Integrated Public Alert and Warning System (IPAWS) capabilities is problematic as more and more of the public relies on cell phones for the majority of their communication. Two—of potentially several—reasons for lack of adoption include system costs for jurisdictions and message originators' education, but even those with access to the IPAWS gateway can be hesitant to use the system. An effort to understand and address barriers to adoption will need to be undertaken.

The social science community's knowledge around public response needs to be exploited when creating messages or using the system. It is important to note the wide diversity in potential message originators. For smaller jurisdictions and organizations, sending alerts may be a part-time job, and a person may only be active in the emergency response community during events; in the largest jurisdictions or organizations, public alerting may be the responsibility of a large team of individuals who are trained emergency management professionals immersed in disaster response full time.

Currently, the rules and standards of practice for issuing emergency alerts are fairly rudimentary. Several organizations contribute to these in differing ways—the Federal Communications Commission (FCC) defines system capabilities, the Federal Emergency Management Agency (FEMA) manages access to the system (requiring training before granting), and private professional organizations, such as the National Emergency Management Association (NEMA), provide some basic education. A framework is needed that provides the following:

- Clear rules of engagement and understanding of the system by local jurisdictions and development of alert and warning templates for smaller jurisdictions;
- Education programs for message originators that focus on understanding mental models in order to shape how these messages will be received, retransmitted (via other channels), and acted upon;
- Published lessons learned or after-action reviews from state and local emergency officials who have used the system for medium to large scale events for review;
- A help desk-style facility either at the federal or state level (see Box 4.1 for an example of similar practice);
- Public education campaigns to increase general understanding of alerting systems;
- Inclusion of private companies in the development of the framework and to encourage ongoing dialogue to explore mutual opportunities for the safety of shared constituents; and

---

BOX 4.1
Contra Costa County's "Rewrite Desk"

Given the challenge to extensively train public safety and emergency management personnel in skills they will only occasionally if ever practice, Contra Costa County, California, by the Office of the Sheriff, utilizes a different approach. Appreciating that there is a great deal of knowledge available in the practice and technology of warning, the sheriff's office maintains an on-call cadre of trained warning specialists who can be temporarily attached to any incident command structure to provide warning support to the incident commander. In addition to mitigating the need to maintain a high level of skill among a large number of people, this "service bureau" (called by one practitioner the "rewrite desk" by analogy to 1930s-era newspaper practice) reduces the uncertainty and anxiety about warning reported by many authorities, largely for the reasons detailed here.

SOURCE: Art Bottrell, California Office of Emergency Services, personal communication.

- Periodic assessments of utility and function, examining new opportunities and technologies. Opportunities are needed to gather researchers to share insights among themselves and with those responsible for defining, building, and operating alerting systems.

Additional understanding on how new systems are integrated and adopted by local emergency management will be important if a truly integrated alert and warning ecosystem is going to be successful. For example, what is the mission and strategic plan for the Wireless Emergency Alerts (WEA) system, or any new alerting system? Does FEMA just enact what the FCC decides or is there a way to continuously modify and update the system? Whose responsibility is that? Should FEMA expand the office and have an operational section and administrative side of WEA/IPAWs? Are there additional roles that NEMA, the International Association of Emergency Managers (IAEM), and other public safety groups could play?

EVER CHANGING TECHNOLOGY

Technology and communications tools used by the public are quite dynamic. The technology itself evolves quickly, and with a growing smartphone application market, the applications used by individuals can quickly change. Alert and warning systems must evolve to make use of these new technologies. This supports the need for a flexible, integrated system.

However, adding to this challenge is that old and new technologies coexist for long periods of time and technologies come in and out of favor. A primary example is broadcast over-the-air television. Although usage of over-the-air television[1] began to decline as cable was adopted, use as shown a slight increase as "cord-cutters" rely more heavily on streaming services.[2] As individuals may still rely on over-the-air television, we must continue to ensure that those individuals receive alerts and warnings. These technology use differences may align with age differences; for example, digital media and streaming services account for 51 percent of

---

[1] A significant amount of work is being done in the United States on the next-generation broadcast television standard, ATSC 3.0. Although the long-term technical and economic viability of ATSC 3.0 is far from certain, it has technical capabilities that could be extremely important to emergency alerting and warning. For more information, see the Advanced Warning and Response Network (AWARN) website at http://awarn.org.

[2] The Nielsen Company, "The Nielsen Total Audience Report: Q2 2016," release date September 26, 2016, http://www.nielsen.com/us/en/insights/reports/2016/the-nielsen-total-audience-report-q2-2016.html.

total average audience for those 18 to 34 years of age.[3] This difference in media choices adds to the complication of reaching differing age populations. While the committee is excited about the possibility of increasingly context-aware alerts and warnings, where the decisions on what type of information to share are made on an end-user device, one must also consider how to most effectively alert individuals who use less advanced cellular handsets, rely on radio broadcasts, or use NOAA Weather Radio All Hazards[4] to get information. Furthermore, both the technology of emergency alerts and citizens' capacity to comprehend the alerts and use messaging functions interactively changes very rapidly. The interaction between the developing technologies and citizens' capacity to use these technologies effectively on a community scale is itself an issue for future research.

COUPLING RESEARCH WITH EMERGENCY MANAGERS AND THE PRIVATE SECTOR

Public response to alerts is both a social science phenomenon (their goal is to change public behavior) and a technical phenomenon (technology is required for their dissemination). It is also an activity closely coupled to the practice of emergency management, which takes place primarily at the state and local level in the United States. Yet technologists (including a wide set of subdisciplines such as human–computer interaction, computer networking, or wireless communications), social science researchers, and emergency managers have had few opportunities for ongoing interactions to consider current knowledge or gaps in our understanding, as well as to collaborate on beginning to fill those gaps. Opportunities and funding are needed for these stakeholders to interact to share their knowledge and experiences and to forge new partnerships.

The Department of Homeland Security (DHS) currently hosts quarterly IPAWS conference calls for emergency managers, which could be expanded to include other stakeholders. Additional research brownbag presentations could be hosted online for researchers and emergency managers to interact. In-person interactions need to be a priority at regularly occurring meetings, such as the International Association of Emergency Managers annual conference and other organized meetings of emergency management. Finally, future funding for WEA research should

---

[3] The Nielsen Company, "The Digital Age, Young Adults Gravitate Toward Digital Devices," release date October 10, 2016, http://www.nielsen.com/us/en/insights/news/2016/the-digital-age-young-adults-gravitate-toward-digital-devices.html.

[4] This text was modified after prepublication.

prioritize integrated teams of technologists, social science researchers, and emergency managers.

INCENTIVES TO PARTICIPATE

An alert and warning ecosystem incorporates numerous official sources of information as well as numerous other information providers, such as social media companies, navigation companies, local media, and hardware makers. For example, WEA relies on cellular service providers to implement the necessary capabilities in their infrastructure and for vendors to include the necessary software in smartphones. (Although participation is voluntary, all major carriers currently participate.) Incorporating these various pieces, and ensuring that information about how the system is working is shared, will be an increasing challenge. How do we encourage openness among stakeholders and encourage participation by those who operate other valuable computer and communications capabilities?

As noted in the previous chapter, feedback on how individuals react to alerts and warnings as they receive them is invaluable to disaster researchers, and companies that might be able to gather this information, like Facebook, Google, or service providers, will need to share this information. Additionally, as private organizations develop their own alert and warning systems, it is essential that they reach out to researchers who have key knowledge on public response, so that lessons learned can be applied (and not rediscovered).

LIMITS IN FORECASTING

Alerting and warnings for naturally occurring events depend on reliably predicting natural phenomenon. Most of the agencies that distribute the messages at the state, local, regional, or federal level rely on the 24-7 coverage of the National Weather Service, the U.S. Geological Survey, and many of the numerous Meso-nets that provide environmental data of all sorts, including weather, earthquakes, and air quality. These agencies must continue to have the resources necessary to provide these vital services that millions of people rely upon. Freely available, high-quality, high-density environmental observations sponsored by the government keep communities, families, and agencies up-to-the-minute with local conditions. The reliance on the public, regular, accurate data for preparedness cannot be overstated. These public networks provide the basis for numerous private networks and will continue to be important even as people use private networks to receive information.

* * *

The continued evolution of alerting systems—and our understanding of the way in which the public uses and responds to these systems—will be essential as an increasing list of disasters and crises happen, both natural and humanmade. Addressing the research enumerated in the previous chapter and careful consideration of the challenges above could be a primary focus of current efforts within the federal agencies that play a role in emergency response, specifically DHS, FCC, and FEMA.

Appendixes

A

Current Alert and Warning Systems and Their Characteristics

The alert and warning ecosystem includes a network of government and industry-supported systems and dissemination channels. A brief history of government-mediated systems is outlined below as well as a discussion of the characteristics of each dissemination channel.

GOVERNMENT-MEDIATED ALERT AND WARNING SYSTEMS

Alert and warning systems have evolved over time to reflect new types of hazards or technologies, moving from various television and radio broadcast technologies to now include cell broadcast. (See Figure A.1 for the evolution of emergency broadcasting.) However, this evolution has occurred very slowly and has often stemmed from a major hazards event. Furthermore, alert and warning systems have not kept up with new technologies. For example, the 2006 Warning, Alert, and Response Network (WARN) Act prompted the first significant changes to national alerting systems since the mid-1990s.

The WARN Act in combination with Executive Order 13407[1] created the Integrated Public Alert and Warning System (IPAWS) and the Wireless Emergency Alerts (WEA) system then known as the Commercial Mobile Alert System (CEAS). The goal of this system was to integrate various alerting systems, the Emergency Alert System, National Warning System, WEA system, and the National Oceanic and Atmospheric Administra-

---

[1] Executive Order. No. 13407, 2006, pp. 1226–1228.

Broadcasting

1951 - 1963
CONELRAD

Originally called the "Key Station System," the CONtrol of ELectromagnetic RADiation (CONELRAD) was established in August 1951.

Participating stations tuned to 640 & 1240 kHz AM and initiated a special sequence and procedure designed to warn citizens.

1963 - 1997
EBS

EBS was initiated to address the nation through audible alerts. It did not allow for targeted messaging.

System upgraded in 1976 to provide for better and more accurate handling of alert receptions.

Originally designed to provide the President with an expeditious method of communicating with the American Public, it was expanded for use during peacetime at state and local levels.

1997 - 2006
EAS

EAS jointly coordinated by the FCC, FEMA and NWS.

Designed for President to speak to American people within 10 minutes.

EAS messages composed of 4 parts:
- Digitally encoded header
- Attention Signal
- Audio Announcement
- Digitally encoded end-of-message marker

2006
IPAWS

IPAWS modernizes and integrates the nation's alert and warning infrastructure.

Integrates new and existing public alert and warning systems and technologies

Provides authorities a broader range of message options and multiple communications pathways

Increases capability to alert and warn communities of all hazards impacting public safety.

FIGURE A.1 Evolution of emergency broadcasting. SOURCE: Federal Emergency Management Agency.

tion (NOAA) Weather Radio All Hazards into one modern network. The new system also looks to take advantage of newer forms of communication, such as cellular telephony, satellite and cable television, and mobile devices. Additionally, IPAWS allows for alerts to be originated by various government organizations and officials at the federal, state, local, and tribal level, and allows a single message to be transmitted to the various alert platforms.

The IPAWS platform simultaneously allows alert originators (AOs) to submit alerts to a server that aggregates and disseminates these alerts to the proper systems. Figure A.2 illustrates the IPAWS architecture. An XML-based data format, Common Alerting Protocol (CAP), normalizes alert data across hazards, jurisdictions, and warning systems. CAP data structure is compatible with existing alert formats but also builds on new capabilities, such as those needed for WEA. IPAWS significantly updated geotargeting capabilities. Older alert systems relied on FIPS codes that were assigned by county or urban area. IPAWS allows flexible geotargeting that can more narrowly target areas using GIS data by entering polygon coordinates to outline the at-risk area. Other new capabilities include multilingual and multiaudience messaging; phased and delayed effective times and expirations, enhanced message update and cancellation features, template support, digital encryption and signature capabilities, and facilities for digital images, audio, and video.

As an example, consider the following emergency situation: A storm produces large amounts of rain, water-level sensors in a local creek in Boulder County indicate rapid onset of a flooding event, and the National Weather Service (NWS) issues a flash flood warning for a polygonal area that the storm will hit hardest. The local emergency management coordinator, who serves as an AO, crafts an alert. The AO submits this alert (in the form of a CAP message that includes the type of event and area impacted) to IPAWS, which then, after authenticating the message, sends it out to the appropriate Emergency Alert System (EAS), WEA, NOAA radio, and other disseminators. In the case of WEA, the cell carriers would use cell broadcast to transmit the alert to a set of cell towers that correspond to the specified area.

Emergency Alert System

EAS is a national warning system originally put in place to enable the president to speak to the United States within 10 minutes and largely used today as a way to alert the public of hazard events. EAS messages are transmitted via AM, FM, broadcast television, cable television, land mobile radio services, and more recently, digital television, satellite television, digital cable, and satellite and digital radio. Messages, which have

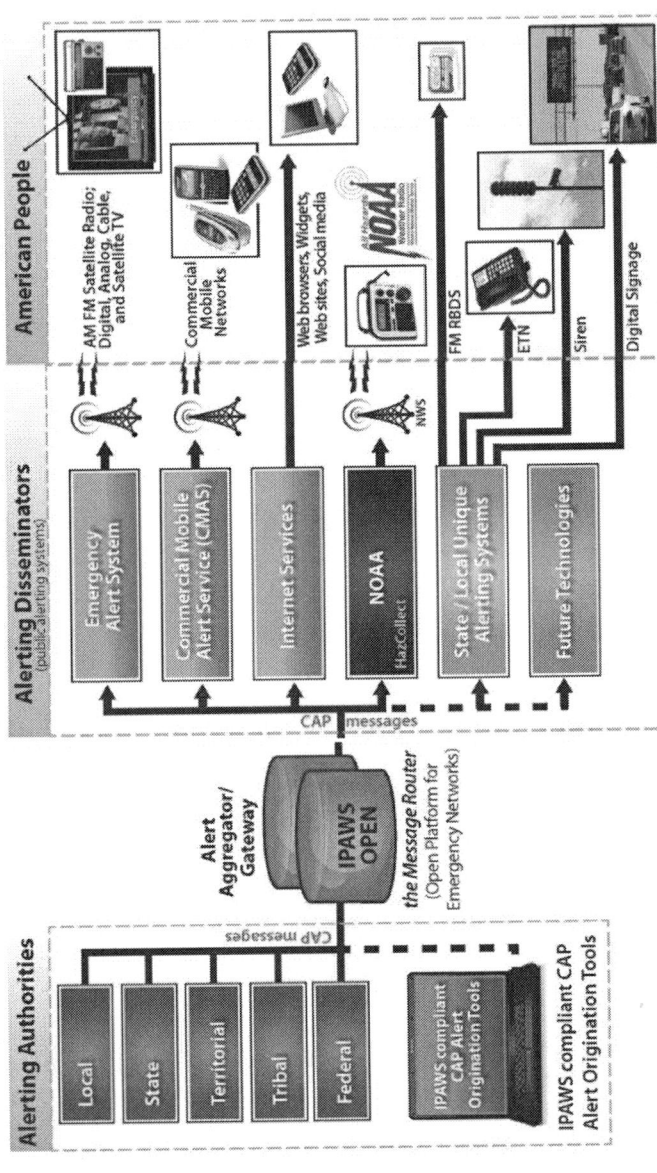

FIGURE A.2 The Integrated Public Alert and Warning System (IPAWS) architecture. NOTE: CAP, Common Alerting Protocol; ETN, Emergency Telephone Notification; NOAA, National Oceanic and Atmospheric Administration; RBDS, Radio Broadcast Data System. SOURCE: Federal Emergency Management Agency. See http://www.fema.gov/pdf/emergency/ipaws/architecture_diagram.pdf.

no length limits, must contain a Specific Area Message Encoding (SAME) header—which contains originator, a short description of the event, date/time issues, and identification of originating station—an attention signal, an audio announcement, and a digitally encoded end-of-message marker. There are currently 80 types of events for the use of EAS.

Wireless Emergency Alerts

As required under the WARN Act, the Commercial Mobile Service Alert (the original name for WEA) Advisory Committee (CMSAAC) was established in late 2006 by the Federal Communications Commission (FCC) to engage stakeholders in the development of initial policy and procedures for one component of that national system—the use of cellular telephones for alerts. CMSAAC, composed of representatives from service providers, handset vendors, emergency personnel, and industry groups, issued its first report in 2007, defining the CMAS's basic system architecture and establishing technical standards and operating procedures.[2]

Messages were restricted in length to 90 characters and the inclusion of URLs was explicitly prohibited. Messages were divided into three categories, presidential alerts, imminent threat alerts, and child abduction alerts. WEA is an opt-out system; cellular customers receive imminent threat alerts and AMBER Alerts, unless they opted out, and cannot opt out of presidential alerts.

Carrier participation is voluntary; to date the major cellular carriers have signed on to the program. WEA-compatible handsets use a special alert tone for WEA messages—to draw attention to the messages and to distinguish them from other messages. This tone overrides the normal ringer-volume settings. A unique vibration cadence is also used to reach hearing-impaired users.

The message format defined for WEA was based on the Common Alerting Protocol.

WEA uses cell broadcast technology known as short message service-cell broadcast (SMS-CB) to transmit messages. Cellular broadcast offers two principal advantages over SMS. First, a single broadcast message can reach each active cell phone within range of a given cellular tower, reducing the network capacity required for message delivery compared to that required for sending messages to each subscriber. Moreover, because

---

[2] The recommendations of the CMSAAC appear in its draft report: Commercial Mobile Service Alert Advisory Committee, 2007, *Commercial Mobile Alert Service Architecture and Requirements*, PMG-0035, FCC, Washington, DC; and in Federal Communications Commission, 2008, Notice of Proposed Rule Making on the Matter of Commercial Mobile Alert System, Public Safety Docket No. 07-287, Washington, DC.

cellular broadcast uses a data channel separate from that used for other messages and calls, it is unaffected by network congestion.

A September 2016 update to the rules for WEA provided significant enhancements,[3] including the following:

- Increase message length to 360 characters for 4G long-term evolution (LTE) and future networks.
- Allow embedded phone numbers and URLs to be included.
- Deliver to a more narrowly defined geographic area.
- Establish a new class of alerts, public safety messages, to convey essential and recommended actions (e.g., emergency shelter locations or boil water order).
- Require providers to support transmission of Spanish-language alerts.

Opt-In Alert and Warning Systems

Jurisdictions have leveraged the ability to deliver messages via SMS and email by purchasing third-party systems that allow for opt-in registration. Emergency management organizations market these systems to the public, who register to receive either SMS or emails about a range of topics. These types of systems are also often used for location-of-interest alerting, for example, school systems and universities; some utilities also use these systems to alert subscribers of system outages. Both SMS-capabilities as well as email only reaches those who register ahead, which limits the reach of these systems.

OTHER ALERT AND WARNING SYSTEMS

Several private organizations have built various alert and warning systems within their own communication networks. These include various weather applications that allow users to receive notifications when weather alerts are issued by the NWS or Google.org's overlay of various alerts and warnings on its maps.

CHARACTERISTICS OF EMERGENCY MESSAGE DISSEMINATION CHANNELS

There is a wide variety of alert and warning dissemination channels available to public message providers in the United States. These are

---

[3] Federal Communications Commission, FCC 16-127, adopted September 29, 2016, https://apps.fcc.gov/edocs_public/attachmatch/FCC-16-127A1.pdf.

listed in Table A.1 and evaluated in terms of the speed at which each can deliver the message, the coverage area that each channel reaches, the degree to which the channel reaches everyone versus only people in focused locations, and the extent to which each channel can provide detailed information in the following table. An inspection of this table illustrates that each communication channel has both advantages and shortcomings. The key shortcoming of WEA messages (limited to 90 characters at the time the table was prepared) is that those messages lacked message comprehensiveness.

TABLE A.1 Characteristics of alert and warning dissemination channels in the United States

Dissemination Channels	Speed[a]	Coverage[b]	Concentration[c]	Message Comprehensiveness[d]
Route alerting	Slow	Limited	Concentrated	High
Loudspeakers and public address (PA) systems	Fast	Limited	Concentrated	Medium
Wireless Emergency Alerts (WEA)	Very Fast	Widespread	Dispersed	Very Low
Wireless communications (SMS)	Very Fast	Widespread	Dispersed	Very Low
Radio	Moderately Fast	Widespread	Dispersed	High to Low
Television broadcast	Moderately Fast	Widespread	Dispersed	Very High to Medium
Television message scrolls	Moderately Fast	Widespread	Dispersed	Low
Newspaper	Very Slow	Widespread	Dispersed	Very High
Dedicated tone alert radios	Very Fast	Limited	Concentrated	High
Tone alert and NOAA Weather Radio	Fast	Widespread	Dispersed	High
Text Telephone (TDD/TTY)	Fast	Widespread	Dispersed	Low
Reverse telephone distribution systems	Fast	Limited	Dispersed	High
Audio sirens and alarms	Fast	Limited	Concentrated	Very Low

[a] The rapidness of the system to reach its targeted audience ranges from Very Fast (less than 10 minutes to Slow (greater than 60 minutes).
[b] Coverage is the size of the area that can be reached by the channel (Widespread, a large area, or Limited, a small area).
[c] Concentration is the degree to which the people that the channel reaches are co-located or dispersed (Concentrated, the message is delivered to targeted locations only or Dispersed, the message has the potential to reach everyone).
[d] Comprehensiveness, or the ability to convey the content needed for effective response classes, used in this table are as follows: Very Low (alerting only); Low (very little information conveyed); Medium (many but not all essential contents conveyed); High (all relevant content conveyed); Very High (all relevant content conveyed with enhanced graphics).
SOURCE: J. Sorensen and D. Mileti. 2014. Protective Action Initiation Time Estimation for Dam Breaches, Controlled Dam Releases, and Levee Breaches or Overtopping. Paper prepared for U.S. Army Corps of Engineers Institute for Water Resources, Risk Management Center. Davis, CA.

B

Summaries of Research Results from DHS-Supported Principal Investigators

The committee asked each of the principal investigators funded by the Department of Homeland Security Science and Technology Directorate to conduct research on emergency alerts to provide a brief summary of their work. These brief statements are included below. Although the committee reviewed these statements, these individually authored papers do not necessarily reflect the views of the committee.

PUBLIC RESPONSE COGNITIVE MODELING OF THE IMPACT OF WIRELESS EMERGENCY ALERTS

Courtney D. Corley, Ph.D., Senior Data Scientist and Team Lead, Pacific Northwest National Laboratory

Wireless Emergency Alerts (WEAs) are a critical mitigation measure employed during emergencies to inform and keep the public safe. Research on WEAs and disasters conducted by the Pacific Northwest National Laboratory (PNNL) and Advanced Brain Monitoring (ABM) has found that individuals perceive the threat of floods differently than other types of disasters on a physiological level within the frontal lobes of the brain. This difference occurs both when subjects are told they are about to watch a video about floods and when they are watching or reading alerts about floods. The perceived urgency of floods also appears to be more sensitive to the personality characteristics of individuals than during other types of disasters.

Methodology

The PNNL and ABM effort collected 20-channel electroencephalography (EEG) data from 51 subjects as part of an experiment to evaluate the ways in which people perceive different kinds of disasters, and their response to different types of social media content related to disasters. Subjects were presented with a series of 50 WEA and Twitter messages collected from each of five types of disasters (blizzard, flood, gas leak, hurricane, and tornado) for a total of 250 messages, and asked after reading each if they would share that message over their own personal social network. These messages were a combination of those shared by actual Twitter users and disaster alerts sent by news stations and other emergency alert services at the time of the disaster. Prior to exposure to a disaster-specific set of messages, subjects were told what type of disaster they were about to view, and then shown a contextual news broadcast related to that type of disaster. All subjects were exposed to the same 50 WEA and Twitter messages for each disaster, but the order in which the disasters were presented was changed randomly each time.

Subjects responding to Wireless Emergency Alerts and social media messages were more predisposed to share WEA and disaster tweets expressing a dismissive sentiment (i.e., a message that advocates or expresses intent to ignore a disaster alert) about floods than they were other types of disasters. Analysis of EEG data from the subjects during the period of time when they were deciding if they would share a given message with their peers over a social network suggests that this decision-making process occurs primarily within the frontal lobe. This is significant because it aligns with other published research postulating the frontal lobes are essential for all aspects of decision-making and play an important role in many higher cognitive functions.[1] Subjects in our study typically had higher levels of brain activity when deciding to share a message as compared to when deciding not to share a message, suggesting a more deliberative thought process. Brain activity changes were especially pronounced when subjects were choosing to share disaster alerts. Older subjects (age 50+) were significantly more likely to share messages of all types with their social network than younger subjects. Overall, all subjects were highly responsive to all types of disaster messages (WEAs and tweets) and shared them a majority of the time.

---

[1] D.A. Pizzagalli, R.J. Sherwood, J.B. Henriques, and R.J. Davidson, 2005, Frontal brain asymmetry and reward responsiveness: A source localization study, *Psychological Science* 16(10):805-813.

Video Response

Our research suggests that subjects have different brain responses toward different types of disasters that are inversely correlated with the volume of danger perceived. During the subject trials, all subjects were shown a context video (news coverage of the specific disaster) immediately prior to responding to WEAs and tweets associated with that disaster. Subject brain activity during these videos was analyzed and compared across disasters types to assess how the subjects perceived the disasters. Previous research by Dennis et al. (2010), exploring the impact of emotional film clips, discovered that subjects with higher levels of electro cortical activity in the frontal lobes were less affected by the stimulus, and the influence of the stimulus was shorter lived.[2] This is consistent with what we observed. Our analysis also found that subjects with the highest levels of activity during the video stimulus were also those who were less likely to share informative WEAs and tweets about the disaster with their peers. The subject's brain activity prior to the presentation of the context videos was also examined. Before the beginning of context videos, subjects were presented with a message explaining that they were about to see a video and tweets about a particular type of disaster. We observe that users have an immediate change in physiological disposition. Their response, therefore, is not shaped by the particulars of the video itself, but only their immediate, visceral disposition toward that type of disaster. This analysis suggests that subjects' response to floods is due to their fundamental perceptions of the dangers of floods, and not the specifics of the scenario. Conversely, upon being told they were about to view a video about tornados, subjects showed unusually high attention—a stark contrast to the response seen toward floods.

Results

Overall, the WEAs tested proved to be highly effective across all disaster types and when compared to other social messages, the WEAs were among the most shared by the test subjects. However, even when subjects chose to share these alerts, the EEG responses to flash-flood specific alerts were distinct from other disasters. When shown context videos for each type of disaster, and particularly for floods, subjects with the least levels of attention and engagement during the video stimulus were also those who were less likely to share informative tweets about the disaster with their peers. Additionally, subjects more frequently shared

---

[2] T.A. Dennis and B. Solomon, 2010, Frontal EEG and emotion regulation: Electrocortical activity in response to emotional film clips is associated with reduced mood induction and attention interference effects, *Biological Psychology* 85(3):456-464.

messages expressing a dismissive sentiment (i.e., a message that advocates or expresses intent to ignore a disaster alert) regarding floods than they were other types of disasters. These responses appeared to be the most exaggerated when among subjects with the least depressive personality types.

Together, this research suggests that the subjects perceived the threat or urgency posed by a flash flood quite differently than other disasters on a physiological level. The response also appears to occur almost instantaneously, suggesting that the response is perhaps reflexive or develops over their lifetime. This response appears to manifest itself in the form of subjects both appearing less mentally engaged with the news coverage of floods, as well as an increased willingness to ignore or actively dismiss the associated weather alerts. There are limitations with the study conducted that are detailed within this report.

Conclusions

The PNNL and ABM team have one primary recommendation and one secondary recommendation for the use of WEAs coming from this research.

Conclusion 1. When compared to tornado, hurricane, gas leak, and blizzard WEAs, flood WEAs are systematically perceived differently in our study group. This leads the PNNL team to suggest that additional attention be directed at communicating the risk of floods to citizens. For example,

- The WEA could focus on stating specific and direct action for recipients.
- Various formulations of WEA could be disseminated specifically for floods as a special case.
- Although geo-targeting of WEA was not in the scope of the PNNL study, providing citizens with location relevant information may further encourage action.
- Education stressing the seriousness or severity of floods and other similarly dismissed disasters might help reduce the public's flippant response to the alerts.
- Users act as a megaphone for disaster alerts in other instances, amplifying the exposure of the alert by repeating its information, particularly for tornados. The WEA program could identify methods to better harness this effect for perpetuating the flood alerts.
- Further understanding is needed of how citizens perceive the risk of disasters in specific regions or of certain cultures. PNNL noted social

media users dismissing specific types of disaster alerts (floods) based on Southern California flash floods. Citizens in other types of disasters in other locations (i.e., Southern United States as opposed to the Northeastern United States) might treat hurricane warnings with similar disregard because they are more common.

Conclusion 2. The results of this study in combination with several recently published reports support the validity of specific neural responses to various types of communications, narratives, and messaging that can accurately predict human behavior in response to these communications. We recommend the implementation of platform technology to routinely screen emergency message form and content using neurophysiological, cognitive, and other measures to add to a database acquired for comparisons and data modeling. This approach would include developing a database of responses from a diversity of people representative of the US population demographics and regions. Data would be uploaded via a cloud-based portal that is easily accessible with a PC and Internet access.

WEA MESSAGES: IMPACT ON PHYSIOLOGICAL, EMOTIONAL, COGNITIVE AND BEHAVIORAL RESPONSES

Deborah Glik, David Eisenman, Kerri Johnson, Mike Prelip, Armen Arevian, Andrea Martel, UCLA Fielding School of Public Health, Geffen Medical School, and Department of Communication

Effective alerts and warnings for disasters protect people and save lives. Over the past decade as mobile communication technologies have become ubiquitous, disaster and emergency messages sent to end users directly have emerged as promising new practices. In particular, short message service (SMS) text message formats have emerged as a modality that is both practical and popular as the majority of the public now use smartphones. In regard to Department of Homeland Security (DHS) wireless emergency alerts (WEAs) these messages are pushed out through commercial mobile carriers to customers who are located geographically near the hazard, and newer smartphones are "WEA enabled."

While the WEA system and other SMS or text based warning systems and messages are coming online rapidly in governmental agencies, universities, and other organizational settings, research about how these systems work has been sparse. We do not have adequate data about how people act and react when they receive WEA messages in real time. This information is key to designing messages that work with the current technology as well as take account of typical human responses to threat messages, otherwise known as the stress response or the "fight or flight" reaction.

The major goal of this research is to test how short WEA disaster-warning messages are processed by recipients. In a series of laboratory experiments in which participants received simulated warning messages on a smartphone, we measured psychophysiological, emotional, cognitive and behavioral responses of recipients. We conducted experiments on a young (18- to 26-year-old) audience who are part of the wired generation, assuming they are adept in regard to mobile device use and literacy, hence representing an audience who should be most likely able to process and use these messages to inform subsequent disaster response actions. Major issues addressed in this study include the following:

- The impact of receiving simulated WEA messages on psychophysiological arousal
- Relative effectiveness of different WEA message lengths (90, 160, 280 characters) and message content
- Behaviors observed among recipients of initial simulated WEA messages
- The role of personal characteristics, emotions, cognitions, and perceptions among recipients of WEA messages
- How physiological arousal, emotions, perception, and behavior interface with the text message and current mobile device technology
- Difference in response received in a social rather than a solitary context

Study methods included a series of social psychological experiments. The study sample was comprised of undergraduate and graduate students between the ages of 18 and 26 who were attending a large urban university. Once recruited into the study, recipients came to a laboratory and were connected with MindWare technology that monitors physiological functioning. They then received WEA messages on a mobile device with either an active shooter or explosion scenario on their campus. Physiological measures were comprised of skin conductance, cardiac activity, and arterial pressure. Personal characteristics, emotions cognitions, perceptions, and behaviors were measured using surveys, observations, and qualitative interviews.

Major conclusions include:

- WEA SMS text messages do have a significant impact on physiological arousal, emotional response, cognitive processing, and behavior.
- The most reliable physiological indicator was SCR (Skin Conductance Response).
- A message of 160 characters is more impactful than a 90-character message, but there is no clear gain with messages that are 280 characters.

- The most effective message length is the amount of characters that can fit onto the mobile device screen of the recipient in the first alert notification. In new phones this is 160 characters; in older phones it is 90 characters.
- Because of the stress response most recipients only read a few words of text before enacting a more general scan of their immediate environment and most do not go into the application itself.
- Moreover, recipients only remember a few key words. Important words or guidance must be articulated at the beginning of the message not buried at the end and must be specific.
- Many recipients clicked off the messages and many did not believe the messages were credible.
- When people were in groups they were more likely to talk to one another about the messages after they received them.
- However, a larger number of people in all study conditions actually did nothing when they received the messages, and people did not click on the embedded URL.

Other conclusions include the following:

- Short concise concrete images and messages of 160 characters are sufficient.
- Use social marketing campaigns and education to teach people what the DHS WEA technology is and how to use it.
- Build a brand that people trust.
- Create pre-event messages templates that are up-to-date and accessible.
- Build capacity for other organizations to build successful disaster and alert warning systems.

RESULTS OF AN INTEGRATED APPROACH TO GEO-TARGET AT-RISK COMMUNITIES AND DEPLOY EFFECTIVE CRISIS COMMUNICATION APPROACHES

Bandana Kar, University of Southern Mississippi

Research Focus

A warning system, essential for risk communication, is comprised of three main components—a detection subsystem (that focuses on detecting and/or predicting location and time of a hazard event), emergency management subsystem (that focuses on determining the threat posed by a hazard, and the necessity to formulate and disseminate alert and warning messages to the public who are at risk from the hazard), and a public response subsystem (PRS) (that focuses on public receipt and understanding of messages, and their responses in the form of preparatory actions).[3,4,5] A number of technologies are used to alert at-risk populations of impending disasters, which include outdoor sirens, Tone Alert Radio (TAR), televisions, and Wireless Emergency Alerts (WEA) that are disseminated via cellular networks to cell phones and other mobile devices as text messages.[6,7] These technologies follow a hierarchical approach such that alerts and warnings are delivered from recognized sources (e.g., National Weather Service) to emergency responders and ultimately to at-risk populations. By contrast, social media that allows creation of social networks and is horizontally integrated across society has become a popular source of risk information before, during, and after an emergency event.[8,9]

Although significant advances have been made to accurately predict a hazard and its impacts, and disseminate risk information to the public, the effectiveness of existing communication technologies is still a research challenge, specifically, from the perspective of the public response subsystem. The PRS is influenced by two main factors: (i) warning messages—message content and style, message source, message delivery technologies; and (ii) message recipient characteristics—social and psychological characteristics of the public. This project focused on examining the effectiveness of the PRS by undertaking the following tasks in the culturally diverse and heavily populated Mississippi Gulf Coast: (i) determine the coverage area of existing alert and warning technologies; (ii) determine perception of the local emergency management agency (EMA) personnel of available warning technologies; (iii) examine the impact of socioeconomic and cultural characteristics of the public, and message format and content on public response to warning technologies and alert messages; and (iv) examine

---

[3] J.H. Sorensen, B. Vogt, and D.S. Mileti, 1987, *Evacuation: An Assessment of Planning and Research*, Oak Ridge, TN: Oak Ridge National Laboratory.

[4] J.T. Grabill and W.M. Simmons, 1998, Toward a critical rhetoric of risk communication: Producing citizens and the role of technical communicators, *Technical Communication Quarterly* 7(4):415-441.

[5] National Research Council, 2012, *Disaster Resilience: A National Imperative*, Washington, DC: The National Academies Press.

[6] J.H. Sorensen, 2000, Hazard warning systems: Review of 20 years of progress, *Natural Hazards Review* 1(2):119-125.

[7] FEMA, "Integrated Public Alert & Warning," last update August 30, 2017, https://www.fema.gov/integrated-public-alert-warning-system.

[8] B.R. Lindsay, 2011, *Social Media and Disasters: Current Uses, Future Options, and Policy Considerations* (R41987), Washington, DC: Congressional Research Service.

[9] B.F. Liu, L. Austin, and Y. Jin, 2011, How publics respond to crisis communication strategies: The interplay of information form and source, *Public Relations Review* 34(4):345-353.

the role of public participation in message preparation on public response to messages.

Methodology

This study was conducted in the three Gulf Coast counties of South Mississippi (Hancock, Harrison, and Jackson). Other than being the landfall location for Hurricane Katrina, the Mississippi Gulf Coast is susceptible to tropical cyclones and coastal flooding events. The high-risk areas of the coastal counties are heavily populated, consist of culturally diverse communities, including Anglo-Americans, African-Americans, and Vietnamese and Hispanic immigrants, and are home to vulnerable population groups, such as immigrant communities composed of older-generation individuals with limited knowledge of English. By conducting this study in a diverse ethnic setting, the purpose was to identify factors influencing use of warning technologies so that policies can be formulated to increase public response to alerts.

A combination of spatial and statistical techniques using both primary and secondary data sets were implemented for each task. Spatial data sets, such as census boundary and socioeconomic information, transportation and hydrologic networks, coast boundary, digital elevation models (DEMs), land use/cover data, and location and spatial coverage data about alert and warning devices were used to model spatial distribution of vulnerable populations, identify physical risk areas, and determine spatial coverage provided by existing alert technologies. Participatory and action-oriented ethnographic surveys were administered in English, Spanish, and Vietnamese to individuals and EMA personnel to collect primary data about the usability and performance of available warning technologies, format and content of messages, and public perceptions and responses to alert messages and technologies.

Research Results

Task 1: Currently, the EMA personnel primarily use the following four technologies to disseminate alert and warning messages to the residents of the Mississippi Gulf Coast - Siren, Radio/NOAA Weather Radio, TV/National Weather Service (NWS), and WEA/Cell Phones. While sirens are located only in two cities—Bay St. Louis (Hancock County) and Gulfport-Biloxi (Harrison County), and provide barely 1.46% of spatial coverage, each of the remaining three technologies provide more than 90% spatial coverage. In fact, cell phone was found to provide 99.4% spatial coverage although the coverage is influenced by spatial variability of signal strength. Despite limited geographic coverage, sirens are still used as their availability is a criterion in the Community Rating System of the National Flood Insurance Program.

Task 2: According to the EMA personnel, in addition to the technologies identified in task 1, Reverse 911, sirens, and posters/pictures are also used to disseminate warnings and alerts. The EMA personnel indicated that TV/NWS, Radio/NOAA Weather Radio, WEA/Cell Phones, and Reverse 911 allow dissemination of updated and accurate warning and alert messages, and therefore, have been found to be more effective in communicating risk information to the residents of the three study counties during a hazard event. Because 90% of the time warnings and alerts are disseminated in English, sirens and posters/pictures are the only technologies available to reach out to the Vietnamese and Hispanic residents with limited English knowledge. However, these technologies fail to provide frequent updated messages to residents.

The EMA personnel also indicated their preference to use social media, specifically Facebook, to disseminate messages during a hazard event. Because social media sites allow public participation and broader outreach, EMA personnel consider this technology to be more effective in motivating the residents to take preparatory actions in response to alert messages. However, due to lack of skilled personnel and the possibility of generating rumors, EMA personnel are reluctant to use social media.

Task 3: Analysis of physical risk revealed that the three counties are occupied by moderate to high risk zones susceptible to coastal flooding. These risk zones are occupied by major urban areas, densely populated, and have a high concentration of socioculturally vulnerable groups.

Analyses of the coastal residents' responses to a household survey revealed that the residents use the technologies identified by the EMA personnel to receive alerts except for posters/pictures that are used by a small group of the population, specifically, the Vietnamese residents. Respondents indicated that they trust information received from their family/friends followed by the information received from authorities via TV, Radio, and WEA/cell phones. Despite limited use, survey participants indicated that they trust sirens more than social media sites. Family and friends were also found to be more effective in encouraging residents to take positive actions as opposed to messages received from authorities. For instance, a smaller, yet sizable, number of respondents indicated that their decision to evacuate will depend upon the actions of family/friends or on their own perception of risk rather than the alerts and warnings received from official sources. However, analyses of survey responses revealed the overall willingness of the residents to follow official evacuation notices.

Almost half of the respondents (130 of 275 survey participants i.e., 47.3%), which included a majority of respondents from Asian and Hispanic ethnicity, expressed their wish to receive alert and warning messages in languages other than English. These individuals indicated that they use social media and family/friends to receive risk information about hazard events, wanted local EMAs to use social media to disseminate alert messages during hazard events, and believed that social media would allow them to receive messages in different languages. The survey respondents indicated that WEA message should include the following information: *nature of the disaster, impact zone, time frame and duration of the disaster, recommended actions, evacuation routes, when to take action, shelter location, how to obtain additional information, and a map of evacuation routes, shelters, and nearby hospitals.*

Task 4: There was a strong agreement (about 60%) among survey respondents regarding (i) their willingness to collaborate with local EMAs in disseminating alerts and warnings; and (ii) their inclination to take positive actions in response to alert and warning messages if they were involved in message dissemination and preparation. These respondents also revealed that they had never participated in message dissemination and preparation with the local EMAs.

Conclusion

The culturally and ethnically diverse communities of the Mississippi Gulf Coast trust and rely on alert and warning messages received from traditional media using conventional technologies. However, these at-risk communities also use social media sites to communicate with families and friends, and have indicated that they would like EMAs to use social media to disseminate risk information. Despite considerable progress in the development and implementation of WEA, its usage remains varied across communities of the study area. Although a majority of respondents claimed to have known about WEA prior to this research, a sizable minority did not, which suggests that more focus is needed to increase awareness about WEA to specific segments of the population. Hispanic and Vietnamese residents tend to use social media during a hazard event for risk information due to the possibility of receiving alerts and warnings in languages other than English. These population groups want EMAs to disseminate alerts in other languages.

Research Gaps and Future Directions

Future research should examine (i) ideal message content for maximum effectiveness, (ii) why some individuals/socioeconomic groups are more inclined to heed messages than others and/or participate in message dissemination than others, (iii) how language barrier affects response to messages, (iv) how EMAs can maximize social media to disseminate messages, (v) to what extent current social media usage provides accurate information to be effectively used by the public, and (vi) how social media usage is impacted by different types of hazards. Studies should also investigate (i) ways to increase public participation in message dissemination process, and in generating relevant and reliable data and information to be used by EMAs during response and recovery activities in near real-time, and (ii) why some individuals and/or socioeconomic groups do not wish to participate in message dissemination nor to take appropriate actions in response to valid alerts and warnings. Actions should also be taken to increase public knowledge of WEA messages, EMAs' social media presence, and ways to disseminate messages to individuals who do not understand English.

COMPREHENSIVE TESTING OF IMMINENT THREAT PUBLIC MESSAGES FOR MOBILE DEVICES

Brooke Liu (PI), Hamilton Bean (Co-PI), Michele Wood (Co-PI), Dennis Mileti (consultant), Jeannette Sutton (researcher), and Stephanie Madden (graduate research assistant)
National Consortium for the Study of Terrorism and Responses to Terrorism

This project sought to determine the optimized message contents of WEA messages. The project compared first-alert WEAs to 140-character and 1,380-character messages. The project also tested 280-character messages should the length of future WEAs be expanded. Research methods were 50 think-out-loud interviews, 13 focus groups, 16 experiments, and 1 community event survey. Hazards examined were active shooter, nuclear explosion, tornado, tsunami, and flash flood. The research resulted in the following conclusions:

- *Message content order.* A different order for the content contained in 90-character WEAs may improve public response outcomes. WEAs currently use the following order: hazard, location, time, guidance, and source. An alternative order had an advantage in improving the public outcomes tested: source, guidance, hazard, location, and time. For 280-character messages, moving the source to the start of a WEA was optimal; however, the optimal order for all WEA content seems to depend on message length.

- *Relative importance of content elements.* For 90-character messages, guidance and hazard message content elements played key roles compared to other message content elements (location, time, and source) in facilitating the sense making outcomes of interpretation (understanding, believing, and deciding) and personalization. They also reduced milling (causing delay in taking a protective action). For 280-character messages, the message elements of guidance (what to do and how to do it) and time until impact (how much time people have to take the recommended action) play major roles.
- *Expanding message content.* Should it become possible to expand the length of WEAs in the future, it would be most important to expand the content areas communicating the hazard (what happened), guidance (what to do about it), and time until impact (when to do it).
- *Familiarity with the WEA service.* Continued outreach and education about the WEA service may help to speed the rate at which members of the public read and respond to WEAs. Survey findings suggest that some people who receive WEAs do not read them immediately. Survey findings further suggest that, when received and read, WEA messages can be effective at reaching and motivating immediate protective action taking.
- *Understanding of acronyms.* The public may have little or no understanding of many of the acronyms used in WEAs. Hence, consideration should be given to modifying the system to discontinue the use of acronyms, educate the public about their meaning, or increase the message length to allow for full text descriptions rather than acronyms. There may be unique exceptions (e.g., NWS)
- *How to best express time.* The way WEAs express time may confuse the public. Currently, WEAs express time by stating when the message expires so that such messages do not persist in perpetuity. However, expressing time this way is confusing and potentially life-threatening. If time is expressed in WEA messages with language about the time a message expires, consideration also should be given to communicating the time a message "begins" to reduce public confusion.
- *How to best express location.* Given the 90-character limit of current WEAs, the phrase "in this area" does not effectively work to communicate who is and who is not located within the risk area. For example, more than a quarter of WEA message recipients from our community event survey did not think that the message was meant for them. Furthermore, each WEA disseminated message that states "in this area" but does not apply to the individual receiving the message may train message receivers that the phrase "in this area" may not apply to them.
- *Understanding of alert and warning concepts.* The public may not understand basic alert and warning concepts. Messages should not rely on the assumption that the public understands terms such as shelter,

evacuate, and proceed to higher ground. For example, survey respondents who reported receiving a WEA message and hearing outdoor warning sirens ranged widely in what they thought proceed to higher ground meant.

- *Map inclusion.* More research is needed on how maps impact public understanding of warnings and differences among hazards. High information map inclusion (specifying the areas affected, areas not affected, and the receiver's location) in 90-character messages had a statistically significant and positive effect on public response outcomes including interpretation and personalization in the case of active shooter and radiological hazards. Inclusion of a low- information map (specifying the areas affected and not affected, but not the receiver's location) had the opposite effect. However, in the experiments testing the 280-character messages for tsunamis none of the map elements tested had a statistically significant effect on message outcomes and research participants varied widely in their reactions to the tested maps.
- *Inclusion of a hyperlinks.* Consideration should be given to including a hyperlink in warning messages of any length. Findings from our community event survey indicated that those who received a message with a hyperlink had a shorter delay (i.e., less milling) before beginning to check media compared to those who did not receive a message with a hyperlink. Delay before avoiding flood areas also was shorter for those who received one or more messages containing a link (compared to those who did not). Focus group research confirmed that including hyperlinks that display additional information may be a useful strategy for expanding the number of characters available for crafting WEA messages, though not all participants supported adding hyperlinks.
- *Message source.* More research is needed on the impact of message source. For 90-character WEAs, findings suggested that local sources may work best except for well-known federal sources like NOAA/NWS. For 280-character messages results indicated that participants had significantly higher levels of message understanding and message deciding when the message came from NOAA than when the message came from a county emergency management agency.
- *Generalizing across hazard types.* Like shorter messages that are 90 and 140 characters; 280-character messages likely do not contain sufficient information to overcome people's pre-alert and warning event perceptions of different hazards based on personal experience, perceived risk, and knowledge, which may or may not match the event they face. Hence, like 90- and 140-character messages, 280-character messages offer less to help effectively manage public protective action-taking than messages that are 1,380 characters.

PUBLIC RESPONSE TO ALERTS AND WARNINGS: OPTIMIZING THE ABILITY OF MESSAGE RECEIPT BY PEOPLE WITH DISABILITIES

Helena Mitchell, Center for Advanced Communications Policy, Georgia Institute of Technology

The Center for Advanced Communications Policy (CACP) at the Georgia Institute of Technology conducted research and development activities to gain a better understanding of how people with disabilities respond to Wireless Emergency Alert (WEA) messages. While these needs were proactively considered during the development of WEA, even today, barriers to access remain. It is critical that impediments to emergency message access be identified, explored, and diminished as the number of people with disabilities and those aging into disabilities increases every day—quickly expanding to nearly 20% of the population. CACP, in support of the DHS/S&T mission and DHS Long Range Broad Agency Announcement 12-7, undertook a number of tasks to address some of these barriers. Project researchers hypothesized that greater awareness and exposure to WEA alerts would increase trust and appropriateness of individual responses to alerts. CACP conducted a national online survey to test the hypothesis and collected data on the availability, awareness, and accessibility of WEAs. The objective was to increase understanding of how to optimize WEA messages, and the devices on which they are received, in a way that encouraged appropriate protective actions. Additionally, numerous development efforts were undertaken, guided by the technical and engineering team. The prototype development was informed by market and user needs analyses. Tasks included testing of WEA-capable handsets, focus groups on experiences with vibration and light features, constructing the prototype, subsequent refinements regarding vibration and light features, and finally the proposed vibration rating (V-rating) scale. The development activity included creating an architectural design for six prototype "handsets" and conducting usability tests with the target population to determine the optimal vibration strengths and the utility of adding a light signal to increase WEA message recognition.

To distinguish the vibration, light and sound qualities within the project, CACP developed the following prototype signals for evaluation:

- WEA Sound. Although evaluating the WEA Sound attention signal was not initially part of the proposed research, we included it in the prototype design to be inclusive of all current and prospective WEA attention signals.
- WEA Light. Adding a WEA light cadence had the ability to increase response time to WEA messages for certain populations. For

participants who were hard of hearing, WEA Light had the quickest response time.
- WEA All. For participants that had low vision the WEA All (i.e., light, sound, and vibration) received the quickest response time over any of the vibration signals alone. Therefore, we concluded that simultaneously activating all notification signals: sound, vibration, and light would increase the likelihood of timely receipt of WEA messages.
- V-Rating. The backbone of this project was to create a vibration rating scale that would be valuable to the FCC, industry, and consumers. The cost to manufacturers to implement this V-rating should be low. Several measurements are already required by the FCC to sell a product and testing the vibration strength of their device could be done and documented at the same time.

Top Conclusions

CACP's research yielded the following conclusions:

- People with prior knowledge of WEA were more likely to take immediate action, less likely to be unsure of what action to take, and less likely to make judgments on whether the emergency alert applied to them. Unfortunately, survey respondents without a disability were twice as likely to report having heard of WEA than those respondents with a disability.
- The survey results showed that 98% of the respondents owned a mobile phone. Regarding wireless only households, 40% of respondents with disabilities reported they did not have a landline phone, suggesting that for those respondents, mobile devices are an important communications tool. Therefore, it is imperative that WEA messages and the devices on which they are displayed be optimized for accessibility with respect to the attention signals and message content.
- Our initial proposed research concerning the vibration and light intended only to target people who were deaf and hard of hearing. However, focus group research incorporating diverse groups of people with sensory disabilities found that people with vision disabilities sometimes relied on the sound and visual attention signals as well.
- Assistive technology device alerting mechanisms and WEA-capable cell phones have a wide range of amplitude, frequency, cadence, and duty cycles in their sound, vibration, and light signals. This inconsistency could impact the user's perception of the WEA vibration cadence (i.e., delayed recognition or none at all). A vibration scale (V-Rating) should be implemented for an end-user to determine the compatibility of a wireless device with their abilities and, therefore, optimize their receipt of WEA

messages. Publishing the V-Rating scale would help manufacturers promote their devices that fall into the range of user needs.
- Data concerning the vibration, sound, and light attention signals indicated a strong need for manufacturers to improve handset effectiveness for both people with and without disabilities. Consequently, if mobile phone manufacturers design handsets with the capability to adjust the strength of the vibration and sound and to include a light feature, both they and the consumer would benefit.

Research Gaps

As WEA has only been active since April 2012, to truly understand its effectiveness, more empirical and case studies need to be conducted. Longitudinal national surveys are still needed. Following are additional research gaps and areas that need further exploration:

- Validate and expand the V-rating scale. It is unclear as to why subjects responded faster to the low and mid-range vibration setting. The vibration motor used in the prototype vibrated at a different frequency for the stronger setting, which could have impacted detection. Further study of the vibration frequency could ensure optimal vibration settings. A frequency scale could then be incorporated into the V-Rating.
- The majority of survey respondents keep their phone in a case, and a small percentage reported that the case affects the phone's signaling to its user. Almost half the respondents reported they "did not know" if the notification signal was affected. More research on this topic is needed to determine the impact.
- The majority of devices that are WEA-capable are smartphones, and a significant number of these have multiple accessibility features. As a result, phone accessibility is diminishing as a barrier to receipt of messages. Nonetheless, efforts should be focused on ensuring that WEA-capable devices are developed to have full out-of-the-box accessibility as findings showed that only 8% were fully accessible.
- Explore aspects of message content to determine how to make WEAs clear and actionable (i.e., encourage more individuals to take immediate protective actions, rather than spend time verifying information contained in the alert).

Future Research

Continued research is needed to inform NG-WEA regulatory rulemakings on accessibility. This research should include:

- Identifying a range of light cadences for WEA messages;
- American Sign Language (ASL) video translation of WEA messages; and
- Exploring technical, and user experience requirements for the integration of symbology and other non-text media into WEA messages.
- Developing and evaluating technology and policy options for the integration of wearable technology into the WEA/IPAWS environment. Wearable technologies are a growing market, and according to our 2015 survey results, both people with and without disabilities have adopted its use at the same rate (14%).
- Determining whether people would be amenable to a feature that turns on the device when a WEA message is detected. Respondents with a disability are significantly less likely than respondents without a disability to keep their phones powered on while sleeping, resulting in missing WEA messages that might be sent during sleeping hours.

OPPORTUNITIES, OPTIONS, AND ENHANCEMENTS FOR THE WIRELESS EMERGENCY ALERTING SERVICE[10]

Martin Griss, Hakan Erdogmus, and Bob Iannucci
CyLab Mobility Research Center, Carnegie Mellon University,
Department of Electrical and Computer Engineering

Research Activity

We consider the essential purpose of the Wireless Emergency Alerting system to be the delivery of the right alert messages to the right recipients at the right time via mobile phones subject to (perceived or real) constraints imposed by the cellular networks. The primary goals of our research are to gain insight into WEA adoption and acceptance issues, in particular with respect to perceived poor public response to alert messages, and to develop and test strategies for overcoming these issues through incremental extension of the WEA service architecture. Our studies involved

- A comprehensive Alert Originator Requirements Study (AORS) to build our understanding of how alert originators (AOs) view and use the WEA service,

---

[10] This research was sponsored by the Department of Homeland Security, Science and Technology Directorate, First Responders Group (Award HSHQDC-14-C-B0016), to advance the WEA Research, Development, Test and Evaluation Program. The full report is available at the DHS web site [1]. Three summary papers are also available [2–4].

- Exploration of a potential connection between WEA and social media,
- Exploration of improved targeting and content. We developed a prototype of a complete enhanced WEA service and then conducted two week-long trials with 225 subjects across three controlled experiments, exploring the usefulness of various features that could be added to WEA.
- Analysis of the feasibility and conceptual costs of extending the current WEA framework to support new features shown to be valuable.

Research Results

Thirteen in-depth interviews with alert originators in local, county, state, and national roles across a range of alert domains and regions of the country yielded these insights:

- Most interviewees were unable to craft meaningful messages to the general population within the constraint of 90 characters.
- A majority of the interviewees stated that WEA needs increased geographic precision to deliver alert messages effectively.
- The WEA service needs to interface with social media to be relevant.
- The nature of WEA is bifurcated: some AOs perceived a WEA message as a "bell ringer" technology while others believed that wireless alerts should directly embed or reference additional information.
- Interviewees agreed that not enough has been done to educate the general population about what WEA messages are, why they are important, and how the public should respond.

These insights drove us to focus on means to improve both targeting and content efficacy of alerts. Targeting is currently based on the assumption that presence in a targeted area indicates interest, and absence indicates disinterest—which may well be false. We call this misconception the location proxy fallacy.

In terms of content efficacy, WEA remains tied to the presumed constraints of 2G cellular networks (limited bandwidth) and an assumption that the mode of delivery (simple SMS-like messaging with simple pop-up per-message alerts) meets user expectations. The reality is that today's LTE networks have far more capabilities, and user expectations have evolved over the last twenty years from text to rich media. On top of this, the AO's task in a complex, large scale disaster will be to send multiple, interrelated alerts reflecting an evolving situation. We refer to the presumed need to keep WEA tied to antiquated network limits as the short message fallacy.

Repeated alerting when the recipient is, in fact, not interested or poor content can lead the recipient to disable WEA on his/her phone. The recipient will then miss future alerts that would have been relevant. We call this the opt-out problem. To address these concerns, we hypothesized and evaluated a variety of new mechanisms. Three of these emerged as showing the greatest potential to improve the WEA service:

- Geo-targeting: Fine-grained, precise geo-targeting based on filtering on the phone rather than in the network and aided by a compressed representation of the target area markedly improved alert relevance to the recipients.[11]
- Location History: Location-history-based filtering combines geo-targeting with interest-based targeting (using content, rather than location, as the determining factor). The use of location history to make alert delivery decisions significantly improved relevance to recipients compared to unfiltered alerts.
- Situation Digest: We developed and evaluated smartphone software capable of digesting streams of structured alert messages and presenting these as consolidated views of the latest information. We measured recipients' situational awareness using their time-dependent, aggregate understanding of the type, recommended action and immediacy of an underlying emergency scenario. Overall, there was a marked improvement in situational awareness compared to the single stream normal WEA view. The feasibility study demonstrated a way to graft this on top of existing WEA.

We believe that recipient-context-specific alert personalization (overturning the location proxy concept) and situational digests (overturning the short message fallacy) may together mitigate opt-out and improve actionability. The interested reader is referred to the full report.[12]

Identified Gaps in Today's WEA

Our study resulted in five major conclusions about WEA:

- Deep integration of location-based context materially improves WEA's value. We found strong AO support for this, and our experimentation and feasibility studies demonstrated efficacy and practicality.

---

[11] A. Jauhri, M. Griss, and H. Erdogmus, 2015, "Small Polygon Compression for Integer Coordinates," paper presented at the Third Conference on Weather Warnings and Communication, June 12, Raleigh, NC, https://ams.confex.com/ams/43BC3WxWarn/webprogram/3WXCOMM.html.

[12] M. Griss, H. Erdogmus, and B. Iannucci, 2015, *Opportunities, Options and Enhancements for the Wireless Emergency Alerting Service*, Washington, DC: Department of Homeland Security.

Our research shows support from AOs and from the experiments for a fundamental change in WEAs focus from issuing alerts to creating awareness,[13] pointing to abilities to digest complex situational information and to support the needs of complex and evolving situations.

- WEA must be well positioned as a peer in the social networking pantheon. WEA serves a critical role as an alarm bell. But the inherent limitations of WEA further suggest that it cannot and will not be the only source to which the public will turn.
- Education, testing and measurement are necessary for WEA's success. Our interviews and surveys revealed the need for education of both AOs themselves and the general public. This feedback correlates with results of previous studies.[14,15] We were surprised to learn how few AOs in our study had actually used WEA. AOs further highlighted the need for regular systematic testing of WEA with the public—not unlike Civil Defense, the EBS, and the EAS.
- Rich media integration into WEA is a question of how, not if. Our research, like that of others, provides support for the integration of rich media into WEA. We believe this arises from the fundamentals of widespread use of smartphones and mature, HTML-based information authoring tools.

Future Research

In our final report, we identify future work including larger scale evaluations, further studies into message compression, the use of URLs and/or images in alerts, and other ideas. Additional concepts include:

- Exploiting the capabilities of smartphones: Re-cast on-phone WEA handling software as an app that can be securely updated and evolved without upgrading the phone per se. This affords continual improvements as user expectations, network capabilities, and current alerting research evolve. Legacy WEA can be maintained in parallel for some years to accommodate the long tail of recipients with 2G phones.

---

[13] B. Iannucci, J. Falcao, H. Erdogmus, M. Griss, and S. Kumar, 2016, "From Alerting to Awareness," paper presented at the 2016 IEEE International Conference on Technologies for Homeland Security, May 12, Waltham, MA, https://ecfsapi.fcc.gov/file/60002085022.pdf.

[14] Carnegie Mellon University, 2014, *Study of Integration Considerations for Wireless Emergency Alerts*, CMU/SEI-2013-SR-016, Pittsburgh, Pa., http://repository.cmu.edu/cgi/viewcontent.cgi?article=1781&context=sei.

[15] M. Wood, H. Bean, B. Liu, and M. Boyd, 2015, Comprehensive Testing of Imminent Threat Public Messages for Mobile Devices: Final Report, *National Consortium for the Study of Terrorism and Responses to Terrorism*.

- Extending communications for resilience and local access: Exploit capabilities of smartphone RF subsystems to allow wireless alerting even when the recipient's carrier network is down. Police and other authorities should have the ability to wirelessly broadcast alerts over their own, hardened systems in a way that smartphones will recognize and display as part of a survivable communications strategy.[16]
- Enriching alert creation: Enable AOs to author rich (HTML-based) content, augment CAP to carry it, and expand WEA to WECAP.[17] Rendering should use the browser mechanisms built into phones. Take advantage of LTE broadcast and remove the short text limit. Digitally sign all alerts.
- Closing the alerting loop: Today, WEA is open loop in the sense that alerts go out and AOs only see the results in terms of the collective actions of the served population. Examine the possibility of closing the loop: provide for recipient responses in future alerting apps (e.g., a button saying "this alert was not relevant for me") to enable deeper studies of alert targeting. The responses need not come back during the emergency but can be trickled back over days following an alert so as to not create inappropriate network load.

GEOTARGETING

Wireless Emergency Alerts in Arbitrary Sized Target Areas: Mobile Location Aware Emergency Notification

Emre Gunduzhan, Applied Physics Lab, Johns Hopkins University;
Bharat Doshi, US Army CERDEC (formerly JHU/APL);
Lotfi Benmohamed, NIST (formerly JHU/APL);
Jay Chang, Applied Physics Lab, Johns Hopkins University;
Osama Farrag, Applied Physics Lab, Johns Hopkins University

The Wireless Emergency Alerts (WEA) service provides the ability to send geographically targeted text alerts to the public. However, the current WEA geotargeting mechanism is limited by the relatively coarse granularity of cellular network sites and numerous benefits of more accurate geotargeting for public alerts and warnings have been identified in earlier studies. The Department of Homeland Security (DHS) Science and

---

[16] B. Iannucci, J. Cali, R. Caney, and S. Kennedy, 2013, "A Survivable Social Network," paper presented at the 2013 IEEE International Conference on Technologies for Homeland Security, Waltham, MA, http://repository.cmu.edu/cgi/viewcontent.cgi?article=1174&context=silicon_valley.

[17] B. Iannucci, J. Falcao, H. Erdogmus, M. Griss, and S. Kumar, 2016, "From Alerting to Awareness," paper presented at the 2016 IEEE International Conference on Technologies for Homeland Security, May 12, Waltham, MA, https://ecfsapi.fcc.gov/file/60002085022.pdf.

Technology Directorate (S&T) has engaged the Johns Hopkins University Applied Physics Laboratory (JHU/APL) to investigate methods that can improve the accuracy of the WEA geotargeting mechanism.

JHU/APL proposed a new WEA geotargeting mechanism, called Arbitrary-Size Location-Aware Targeting (ASLAT), and conducted several analyses to characterize the performance of the new mechanism and to assess feasibility of its deployment. JHU/APL compared the performance of the new mechanism with existing WEA geotargeting, developed functional requirements for the new mechanism and identified the required changes to existing WEA and applicable geolocation standards.

ASLAT utilizes the location awareness of mobile devices to improve geotargeting accuracy. In ASLAT, WEA alerts are broadcast to an area wider than the target area, but are only displayed to the user if the mobile device is inside the target area. This approach eliminates the false positives and the false negatives that occur due to the mismatch between the shape of the target area and the shape of the set of cellular network sites selected to broadcast the alert. In addition to enhancing the geotargeting accuracy, ASLAT would enable people to receive alerts when they are in the vicinity of a target area and have interest in a particular location inside the target area.

Performance analysis of ASLAT shows that it can improve the geotargeting accuracy of WEA significantly without consuming excessive mobile device power or radio resources. ASLAT introduces some delay in delivering alerts because mobile devices need to learn their location before processing a received alert. However, the maximum delay introduced by ASLAT can be controlled by a configurable parameter. The mobile device can use the default WEA behavior if the ASLAT delay reaches this maximum value. We used a maximum ASLAT delay of one minute in the analyses. Highly delay-sensitive alerts such as earthquake warnings would bypass ASLAT automatically and be processed using the default WEA behavior. These alerts would be displayed to the user immediately as in current WEA, without comparing the location of the mobile device with the target area.

ASLAT depends on a variety of geolocation technologies to determine the location of a mobile device. We investigated different geolocation technologies to see what technologies would be suitable for use with ASLAT and concluded that mobile-device-based technologies should be used. These geolocation technologies include the Global Positioning System (GPS), mobile-device-based Time Of Arrivals (TOA) and Time Difference Of Arrivals (TDOA) techniques and Wi-Fi proximity. These are all suitable for ASLAT because they provide adequate location precision, they do not introduce additional load on the cellular network and they maintain user privacy.

ASLAT would require some changes to existing standards. Specifically, WEA standards that specify cellular network functionality and mobile device behavior would require amendments to support ASLAT. Some modifications to GPS and TOA/TDOA standards would further enhance ASLAT performance. Since these standards have been developed primarily for navigation and E911 services, extensions to support emergency alerting and the impact of these extensions on existing services need further study. Finally, modifications to Wi-Fi standards and implementation of new indoor location capabilities are needed to enhance geotargeting accuracy indoors.

The proposed geotargeting mechanism and the related results could affect important technical, programmatic and policy decisions regarding the evolution of the WEA system. The DHS S&T WEA Program Management Office should work with other stakeholders, including the Federal Communications Commission, Federal Emergency Management Agency, cellular service providers, the Alert Originator community and state and local first responders, to determine detailed requirements on geotargeting accuracy and to further analyze various alternatives to meet these requirements.

Geo-Targeting Performance (GTP) of Wireless Emergency Alerts (WEA)

Dan Gonzales, RAND Corporation

The objectives of this study were to evaluate the public benefit and performance trade-offs of geo-targeted WEA messages using alternative WEA antenna selection methods and to identify the optimal WEA radio frequency geo-targeted areas for imminent threat scenarios. This briefing addresses these questions for two imminent threat scenarios: tornado warnings and earthquake early warning. The briefing concludes with the following recommendations.

Employ WEA Antenna Selection Method 2 In Urban and Mixed Areas

We found WEA Method 1 provides lower over alerting rates (OARs) than Method 2 in tornado warnings. However, if Alert Failure Rate (AFR) is considered to be the primary metric used to assess WEA GTP, Method 2 provides superior GTP in urban and mixed areas. Because of RF spillover effects it is not clear which WEA antenna selection method is better in rural areas.

Upgrade Sirens and the WEA Service to Improve Geo-Targeting of Siren Tornado Warnings

Previous studies found that siren-based warnings were ignored by residents under threat because of alert complacency or fatigue. Complacency occurred because tornado warning sirens were sounded on a countywide basis in many areas. WEA can be used to geo-target siren tornado warnings. A WEA compatible receivers should be installed on sirens.

Explore the Implications of the FACETs and Threats in Motion (TIM) Initiatives for WEA

Forecasting a Continuum of Environmental Threats (FACETs) is an NWS initiative to improve the NWS forecast and warnings for high-impact weather events. FACETS messages include nested, color-coded polygons to indicate the level of threat present in each area. The current version of the WEA service cannot support FACETS.

Another NWS initiative is Threats In Motion (TIM). TIM will improve tornado warning geo-targeting by updating the position of the warning polygon more rapidly. Today tornado warning polygons may remain fixed for hours. TIM warning polygons are updated every minute. TIM presents a number of challenges for the WEA service. TIM would require the transmission of more WEA messages. The capacity of the IPAWS aggregator will likely have to be increased to support this increased message load. Testing will be needed to ensure WEA can handle TIM-based tornado warnings.

The Federal Communications Commission (FCC) and DHS are considering upgrades to WEA. Such upgrades should consider the implications of FACETS and TIM based tornado warnings. The FCC and DHS should consider changes to WEA that will enable FACETS tornado warnings to one day be transmitted as WEA messages.

WEA Testing Is Needed to Determine Whether WEA Preserves Tornado Warning Lead Time and Can Support Earthquake Early Warning (EEW)

The average lead time or warning time provided by NWS tornado warnings is about 13 minutes. While there appears to be no formal WEA message latency requirement, previous industry studies indicate WEA message latency may be as high as 12 minutes. If WEA tornado warnings are delayed by this much then almost all of the lead time provided by NWS tornado warnings would be consumed by time delays within the WEA service infrastructure.

An EEW sensor network is under development in California. However, a means has yet to be identified for transmitting EEW messages to the public using cell phones. An EEW system could provide 40-60 seconds of warning prior to the start of the worst shaking, which could save many lives in a large earthquake. To provide such warning the EEW message would have to be disseminated in 10 seconds or less to be effective.

A recent industry study determined that the current version of WEA could not support EEW because of WEA message time delays, without presenting specific evidence of this assertion. The timeliness of the WEA service has never been evaluated in an end-to-end test. Such testing is needed to determine how effective WEA tornado warnings are and whether WEA EEW is feasible. Cell broadcast-based warning systems in other countries may have much lower message latency than the U.S. WEA service. If it is found that WEA service message latency is high, it should be technically feasible to reduce these time delays. WEA testing can determine if this is necessary.

DHS or NWS Should Conduct an Education Campaign to Inform the Public that WEA Geo-Targeting Is More Accurate than Sirens

Previous studies indicate that a large percentage of the population ignores siren tornado warnings because of over alerting. Many people may not be aware that WEA tornado warnings can be geo-targeted more precisely than sirens, so they may also ignore WEA tornado warnings. To prevent this an education campaign is required to inform the public of the superior geo-targeting performance of the WEA service.

Develop Tools to Help Alert Originators Estimate WEA Local Area Coverage

The coverage provided by wireless cellular networks can vary significantly from one region to another. This is especially true in rural areas where cell towers are sparsely distributed over the terrain. In contrast in urban areas cellular coverage is generally good and cell sizes are small, which leads to high WEA geo-targeting accuracy. Alert originators in rural areas may therefore have greater uncertainty as to how far a WEA message will propagate and where to draw a warning polygon in an imminent threat scenario. New tools for alert originators that provide WEA coverage estimates would be valuable in such environments.

Exploring the Effect of the Diffusion of Geo-Targeted Emergency Alerts: The Application of Agent-Based Modeling to Understanding the Spread of Messages from the Wireless Emergency Alerts (WEA) System

Andrew M. Parker, Brian A. Jackson, Angel Martinez, Ricardo Sanchez, Shoshana R. Shelton, & Jan Osburg, Homeland Security and Defense Center, RAND

The complexity of individual behaviors in emergency alerting situations, and the way those behaviors affect the ability of alerts to serve as a protective measure, has long been a topic of interest. With the evolution of new relationships between citizens and technology—which have affected the utility of legacy alerting modes and the arrival of new options like WEA—new facets have been added to that complexity. The ability to geo-target alerts to mobile devices provides new capability, but it also raises questions about how to use geo-targeting effectively and how the interaction between targeted messages and human communication behavior will affect diffusion of messages beyond the area where they are delivered initially. This project took on the question of how important diffusion behavior was for understanding the value of geo-targeting WEA messages. The study used agent-based models (ABMs) to examine the diffusion of alerts within an area populated by individual recipients who move from place to place in the course of their daily activities.

Building a Model of Emergency Alerting and Diffusion

An ABM is a computer simulation where a population is represented by individual "agents" (in our model, representing people in at-risk areas) who interact and make individual choices about their behavior based on a set of rules. In our model, the agents could move geographically, since understanding the effect of individuals moving in and out of areas where geo-targeted alerts were sent was the central aim of our work. The model was represented a chain of events, including (1) receipt and understanding of an alert message, whether from WEA or another alerting channel, (2) the decision by the agent to share or not share that alert with others, the central mechanism for alert diffusion in the model, and (3) the decision by the agent whether or not to take whatever protective action was recommended. Agents resided and moved in a realistic geographic landscape, including a network of roads.

We considered four alerting scenarios: (a) flash flood, a frequent focus of past WEA messages with a small area, short timeline, and evacuation goal, (b) tornado, also a frequent WEA focus but covering a larger area with a shelter-in-place goal, (c) hazmat plume, a small-sized event with the potential for multiple hazard areas with differentiated protective actions (evacuation or shelter), and (d) major flood, a rarer but larger and slower event, with the potential for multi-stage evacuation strategies.

By distilling the elements of alerting to a set of simplified parameters, the model allowed us to look at how changing some of those parameters could affect outcomes—dialing up and down the parameters that shaped how people communicated and forwarded messages to get insights into how that behavior could become a multiplier for alerting even as it fought efforts to precisely geo-target alerts.

Key Conclusions from the Model

Our initial framing of the study considered forwarding as a potential threat to the value of geo-targeting. This perspective considers forwarding as producing a practical limit on geo-targeting precision. The simulations certainly showed this—as would be expected—but the relationship was more nuanced. Very tight geo-targeting will always be compromised some by forwarding, but if very few agents are in the alert zone (i.e., are in or near the emergency zone) that compromise will be small, and hence should not hurt the value of investing in geo-targeting. Diffusion is likely of greatest concern for intermediate-sized events, where the population to be alerted is reasonably large, yet the geographic area is small enough to suggest value in precise geo-targeting. As the number of people being alerted in our simulations went up, the number of potential forwarders also went up, and the ability to target precisely went down—though, depending on the scenario (e.g., how quickly it evolved), that could matter to differing extents.

That said, instead of considering forwarding as a threat to geo-targeting, it might be more valuable to think of forwarding as a compliance enhancer. Significant forwarding could increase the total effect of alerting a great deal. In that sense, forwarding converts the set of individuals' social networks into a new mass alerting channel of its own. In this way, forwarding or communication among individuals about the alert, here via such electronic means as social media, is simply one more type of the communication that has always occurred during milling before citizens make the decision to comply with the alert. The price of this effect is significant increases in out- of-emergency-zone alerting and potentially unnecessary action in response.

The model also taught us that precise geo-targeting can have different meanings when we consider how agents are dynamic, rather than thinking about geo-targeted alerts being aimed at a set of agents that is sitting still "waiting for the alert to arrive." We may want to get the message not only to people who are in the emergency zone but also to those contemplating entering the emergency zone.

Our results demonstrate quite clearly that forwarding threatens the value of trying to deliver different messages to different geographic areas in an effort to either provide messages relevant to individual risk areas or to guide population behavior in ways designed to enable more-effective response (i.e., time-phased evacuation). The dynamics of forwarding and compliance mean that success in these sorts of differential alerting efforts would require more-nuanced communication strategies (e.g., telling later staged-evacuation zones to "shelter for now").

Limitations

Perhaps the most important caution to be placed on these results derives from the very nature of such a modeling exercise, which by design is an abstraction from reality. All models are abstractions, and aspects of reality may not have been completely captured (e.g., the current model does not capture non-verbatim forwarding of alerts). As an abstraction from reality, it is important to remain clear that the model itself relies on the estimates of the parameters that serve as its basis, as well as the scenario specifications.

Conclusions and Future Directions

Complexities of human behavior—including both message forwarding and movement—mean that the use of geo-targeting is not as simple as just restricting the transmission of an alert to the smallest area at risk from an emergency event. The ability to transmit messages to smaller areas—which is indeed a major technical jump in emergency alerting capability—requires similar innovation in policy and practice to ensure that emergency managers are outfitted with the best understanding and tools to make the best choices regarding the use of geo-targeted alerts during the critical and time-limited decisions made in the warning phase of natural, technological, or other emergency incidents.

This project demonstrated the value of agent-based simulation models for capturing key complexities—such a social networks and geographic movement—in ways that wouldn't have been possible otherwise. These or similar models could be used to run experiments examining policy questions, such as the strategic use of forwarding as an alerting channel, conditions under which over-alerting pays dividends, and the impact of new WEA capabilities or new technologies.

Using RF Coverage to Improve Geotargeting Granularity and Accuracy for Delivery of WEA

Dara Ung, Comtech Telecommunications Corporation

Comtech TCS carried out research and development on constructing geo-targeting algorithm that utilizes Radio Frequency (RF) cell site propagation footprints. Research focused on using RF coverage area footprints to improve geo-targeting granularity and accuracy for delivery of Wireless Emergency Alert (WEA) messages. The project consists of field testing activities, conclusions, and the analysis of the results that used the enhanced geo-targeting algorithm previously developed for this research.

The WEA standard (J-STD-101) defines two methods that can be used to select cell towers to deliver WEA messages for a given targeted geographical area. The first method calls for the ability to determine the cell towers at the county level of granularity. This level of granularity is a minimum requirement for all mobile carriers that offer the WEA service. The second method is optional and allows the targeted area to be defined by polygons instead of fixed county boundaries and determines if the targeted cell tower physical position (latitude/longitude) is found inside the target area polygon. Both of these methods have been found to be highly inaccurate as the alert target areas become smaller and therefore cannot be used to issue alerts that require target area size to be within a few square miles. This inaccuracy introduces situations known as "over-alert," when an alert reaches population that is not intended for, or "under-alert," when the alert does not reach the people in harm's way.

This research included the modification of WEA software using enhanced geo-targeting algorithms that take into account more than just the physical location of cell towers. The algorithm was tested both in the laboratory and in the live production environment. The outcomes of the research include obtaining the live test results that are keys to validate the lab simulation. The results will also confirm the successful development of tools and software needed to collect the data in the live environment without impacting (due to sending test alerts) to the public.

The test results obtained from the field clearly demonstrated the strength and weakness of both the existing and the new enhanced methods. The results show that the enhanced algorithm using cell RF propagation footprints is convincingly superior to any existing method used today. When implemented, the enhanced method developed in this research will provide new benefits to WEA users in several ways, including:

- The ability to target much smaller alert areas down to a square mile regardless of the physical location of the cell towers;

APPENDIX B 121

- The ability to use location based Required Monthly Test (RMT), allowing WEA alerts to be tested at chosen live site without impacting the general public;
- Geo-targeting at the cell sector granularity;
- Enhanced reachability to the people in harm's way;
- The ability to enable other alert categories to be defined because of allowable small alert target area size; and
- A solution that requires no change to the current WEA network.

The key lessons learned in this project consist of understanding the effects of a live environment and how different real-world factors can affect the expected results. These lessons learned are very important because they allow WEA vendors to improve the techniques that will ultimately enhance their solution in the future.

Due to the limitation of the cell broadcast technology, no geo-targeting method can provide 100 percent accuracy. Based on the results obtained, however, the algorithm that uses cell tower RF propagation footprint clearly offers better accuracy than the methods used to date. Although this method will not solve the over-alerting problem within the cell sector level, it will improve the reachability to people in harm's way very effectively. Since the alert target area size can now be defined as small as a square mile, over-alerting can significantly be reduced. Therefore, this method will be suitable for such alerts as a campus emergency, a chemical spill or a road block due to a major accident. These instances would not be possible using the various methods available today.

The cell RF propagation footprint algorithm could be provided as the best-effort solution for cell broadcast technology currently available. The attractiveness of this method is that it does not require any change in the standards and specifications for it to be deployed today. The existing WEA regulatory mandatory requirement for geo-targeting is limited to county-level only. It is therefore recommended that the regulatory requirement be changed to obligate the service providers to offer WEA service with geo-targeting at cell sector level accuracy.

Given the limitations discussed in this document, further improvement can still be made, perhaps in cooperation with a mobile device application developers or manufacturers. The Dynamic Plus method allows a very small area to be targeted and can identify exactly the list of cell sectors affected by this area. Cell sectors, however, can be quite large and can extend for several miles in rural areas. Over-alerting can therefore extend for miles. To avoid this problem, intelligence in the mobile device is needed. A GPS capable mobile device knows its own location. Thus, if the target area LAT/LONs or circle can be conveyed to the mobile device over the cell broadcast message, the mobile device would be able to determine if it is located inside the target area and subsequently notify the user. Otherwise no alert will be triggered on the device. Such combined technology would provide the best possible result and is recommended as a subject of future experiment.

TECHNOLOGIES

Accessible Common Alerting Protocol Radio Data System Demonstration: Gulf Coast States

Rich Rarey, NPR Labs

The goal of this project was to create and demonstrate end-to-end accessible radio emergency alerting using Common Alerting Protocol (CAP) messages from the FEMA's Integrated Public Alert and Warning System (IPAWS) aggregator. The resulting alert messages were specifically formatted for and targeted to deaf and hard of hearing individuals living in the U.S. Gulf Coast. It was understood that deaf and hard-of-hearing residents living in the weather-vulnerable Gulf States region would benefit directly from receipt of emergency messages that are currently provided in an audio format to hearing individuals.

NPR proposed using existing satellite delivery and broadcast technologies for secure and timely transmission of text alerts in an innovative way: Create a delivery system for emergency information relying on broadcast radio technology that is accessible and available regardless of power outages, Internet disruptions or limitations of cellular service.

Tools created during the project were designed to be used by Public Radio Satellite System (PRSS)-connected FM radio stations and to have broad application for adoption by other commercial and non-commercial FM broadcasters using Radio Data Systems (RDS). RDS is a long-established data subchannel technology, typically used by an FM station to broadcast multiple data streams containing traffic data, artist/song/album data, and similar information.

NPR Labs used the PRSS distribution path to reach 26 public radio stations scattered across Alabama, Florida, Louisiana, Mississippi and Texas. The stations locally broadcast the alerts to deaf and hard-of-hearing participants who received the alerts on specially designed 'visual' FM radio receiver. The disseminated emergency messages from FEMA's test platform, using the Emergency Alert System (EAS) architecture, demonstrated how attributes of the CAP message, embedded within an RDS signal, can greatly expand the distribution of EAS alerts for the nation's 36 million citizens who have hearing loss.

To build the demonstration, significant hardware and software development was required, as well as system integration to provide a way to access, process, and transmit text emergency messages. Vendors were identified and recruited to help design and manufacture a specialized message encoder for the stations, and mass-produce an easy-to-use, specialized 'visual' FM radio receiver for the deaf and hard of hearing users. Additionally, software was developed that automated the relay of the IPAWS alerts from reception at NPR, through the PRSS satellite system, through the stations, to the users, thence displayed to the deaf and hard of hearing participants.

Two rounds of consumer field tests were conducted during the summer of 2014. Sample messages were sent from the PRSS Network Operations Center to local public radio stations for broadcast to field-test participants. Consumers read the messages, some up to 4,000 characters in length, on an Android tablet running an application developed for the project that linked the tablet to the specialized FM-radio receiver. The FM receiver automatically tuned to the strongest signal of a participating station and was designed to be operated by battery and/or wall power. It had a connector to activate a bed shaker to awaken the user. During an alert, bright lights on the FM receiver and the bed shaker notified the user that a message was being received and the text of the alert, on the Android tablet screen, advised the reader of the emergency and actions to be taken. Participants were surveyed daily for their impressions and experiences. Results from the first set of tests were used to improve and upgrade software for the second round of tests.

The project employed best practices, and identified areas in technology development and design, engineering management, station installation, recruitment and management of field-test participants that could greatly improve future product development and testing methodologies. The specialized FM radio receiver developed for this project, dubbed the NIPPER ONE, was a 2014 International Consumer Electronics Show Innovations Design and Engineering Award Honoree. The project work was featured by the White House at its Innovation for Disaster Response and Recovery Day in July 2014. The novel design of the RDS Emergency Alert signaling was incorporated into the National Radio Systems Committee standard *NRSC-G300-A Radio Data Systems (RDS) Usage* (April 2014) and the subsequent update NRSC-G300-B (September 2014).

Recommendations and next steps are in two categories: Refining technological understanding and delivering non-English text alerts. Most critical task in the first category is developing a terrain-based prediction model for RDS coverage (including FM transmission injection level variables) to more accurately predict where and how well users will receive alerts. Also required are enhanced error correction of incoming messages;

improving the look, feel of the consumer hardware while reducing cost; and reducing the cost of station based equipment.

While this project made English language text alerts available, follow-on work could focus on investigating and developing methods to create emergency alerts for non-English speakers. Similar technology could be utilized to deliver alerts in other languages, such as Spanish, Vietnamese, and Chinese.

SEI Wireless Emergency Alerts (WEA) Research 2013 Through 2016

Carol Woody, Software Engineering Institute

Research Activities

The Software Engineering Institute (SEI), a federally funded research center sponsored by the Department of Defense, addressed a research need from Denis Gusty, Department of Homeland Security Science & Technology Division, to assist alert originators (AOs) in their efforts to implement Wireless Emergency Alerts (WEA). SEI developed an integration strategy to aid AOs in adopting and utilizing (WEA). In addition, best practices for addressing requirements, vendor product selection and acquisition, technology choices such as cloud technology, cybersecurity risks, testing, and implementation were assembled and published. Twenty-eight emergency organizations were contacted and fourteen vendors were interviewed. Workshops were conducted to identify concerns and develop feasible approaches. In-depth analyses focused on issues of trust and cybersecurity risk were conducted. The papers were published in the following areas:

- *Integration.* Commercial Mobile Alert Service (CMAS) Alerting Pipeline Taxonomy and Study of Integration Strategy Considerations for Wireless Emergency Alerts
- *Best Practices.* Best Practices in Wireless Emergency Alerts; Wireless Emergency Alerts: New York City Demonstration; Maximizing Trust in the Wireless Emergency Alerts (WEA) Service; and Wireless Emergency Alerts: Trust Model Technical Report
- *Security.* WEA Cybersecurity Risk Management Strategy (for AOs); WEA Security Pocket Guide (delivered as a section of best practices); Mapping WEA Security Requirements and Guidance to Cybersecurity Risk Mitigation Recommendations (delivered separately to DHS - some requirements are restricted); and INCOSE Insight essay, Evaluation of Security Risk for WEA Alert Originators Using Mission Threads (published July 2013, Volume 16, Issue 2)

APPENDIX B

In a follow on project, SEI assessed Commercial Mobile Service Providers (CMSPs) cybersecurity risks that affect the WEA service and developed the Wireless Emergency Alerts CMSP Cybersecurity Guidelines which were published.

Research Results for Alert Originators

Research completed by SEI provides the following advice for alert originators:

- *Requirements.* Spend the time to identify the key requirements, and specify them meaningfully including what the system must do (functional requirements) and how well the system must operate (quality attribute requirements).
- *Cloud trends.* Know what quality of service you need, and ask how the vendor will achieve it; know how to look beyond the jargon and hype.
- *Cybersecurity.* Learn about the security risks associated with modern alerting technologies and establish a culture of good security; Implement a cybersecurity risk management strategy.
- *Product selection:* Confirm WEA capabilities before purchasing the product; develop a customized prioritization method that documents the progression from operational expectations to prioritized features; those that lead tradeoff discussions should acquire sufficient knowledge of tradeoff definitions and consequences to lead these design discussions; address the role of a "lead integrator" if you will use multiple vendor products.
- *Testing.* Attend FEMA IPAWS webinars and outreach sessions; conduct periodic tests of the individual system and software to include interface testing between the alerting software and IPAWS-OPEN using available testing platforms; periodically conduct end-to-end testing using available testing platforms.
- *Operations.* AOs need a method for identifying the operational impacts of WEA; they need to ad-dress and manage operational challenges prior to an emergency incident; Good practices can assist in sending rapid, clear, and timely messages; large-scale exercises and training are important to exercise cross-agency and cross-system scenarios; there are cross-organization coordination challenges in issuing WEA messages; there are challenges in synchronizing WEA information with other media channels.
- *Alternatives to buying a WEA solution.* There are options for obtaining a WEA solution, and each has advantages and disadvantages; there are special considerations for developing your own WEA solution; there are important considerations related to authentication and message validation; there are challenges with error-handling message propagation.

Challenges for Trust and Cybersecurity

Alert originators do not have a security mindset at the leadership and management level; security is seen as "someone else's job." CMSPs have taken steps to address cyber security concerns and the guidelines provide a means for them to review their current security position, identifying gaps, and chart a course for improvement.

WEA, like all other cyber-enabled systems, is subject to technology weaknesses and cyber threats that may prevent its use or damage the credibility of the service that it provides. Attackers may attempt to delay, destroy, or modify alerts, or even to insert false alerts. These actions may pose a significant risk to the public.

Trust is a key factor in the effectiveness of the WEA service. AOs must trust WEA to deliver alerts to the public in an accurate and timely manner. Alert recipients must also trust the WEA service before they will act on the alerts that they receive. All participants in the WEA Pipeline must work together to ensure trust is maintained.

C

Briefers to the Committee

AUGUST 9-10, 2016

Greg Cooke, Federal Communications Commission
Harry Evans, Austin, TX Fire Department (retired)
Karen Fregberg, University of Louisville
Benjamin Krakauer, NYC Office of Emergency Management
Mark Lucero, Federal Emergency Management Agency
Paul Lupe, Fairfax County, VA Office of Emergency Management
Francisco Sanchez, Harris County, TX
Tim Schott, National Weather Service
Matthew Seegar, Wayne State University
Tim Sellnow, University of Central Florida
R.C. Smith, El Paso County, CO Office of Emergency Management
John Sorenson, Oak Ridge National Lab (retired)
Jeannette Sutton, University of Kentucky
Ben Zhao, University of California, Santa Barbara

SEPTEMBER 1, 2016

Court Corley, Pacific NW National Laboratory
Deborah Glik, University of California, LA
Dan Gonzales, RAND Corporation
Emre Gundunzhan, Johns Hopkins University of Applied Physics Lab
Bob Iannucci, Carnegie Melon University
Bandana Kar, University of Southern Mississippi

Brooke Liu, National Consortium for the Study of Terrorism
Helena Mitchell, Georgia Tech Applied Research Corporation
Andrew Parker, RAND Corporation
Rich Rarey, Rareworks, LLC
Dara Ung, TeleCommunication Systems, Inc
Carol Woody, Carnegie Melon University Software Engineering Institute

NOVEMBER 1-2, 2016

Art Botterell, California Governor's Office of Emergency Services
Peter Cottle, Facebook
Nelson Daza, Everbridge
Julie Demuth, National Center for Atmospheric Research
Pete Giencke, Google
Alex Kalicki, Facebook
Jim Moffitt, Twitter
Micah Schaffer, Snapchat
Deanna Sellnow, University of Central Florida
Michele Wood, California State University, Fullerton

JANUARY 26-27, 2017

Dave Bujak, Weather STEM
Brian Daly, AT&T
Farrokh Khatibi, Qualcomm
Karl Kotalik, NC4
John Lawson, AWARN
Christopher McIntosh, ESRI
Brenda Phillips, University of Massachusetts, Amherst

MARCH 23, 2017 (TELECONFERENCE)

Maiyan Bino, Waze
Rebecca Resnick, Waze